Sydney's Bushland

More than meets the eye

Jocelyn Howell & Doug Benson

Principal photography by Jaime Plaza

Royal Botanic Gardens Sydney

SYDNEY • DOMAIN • MOUNT ANNAN • MOUNT TOMAH • NATIONAL HERBARIUM

Acknowledgements

It is impossible to name all who have contributed over the years to our understanding of Sydney's bushland, but we would like to thank them. Among those who have helped more specifically with producing this book, we would particularly like to thank Penny Farrant and Helen Stevenson for their cheerful cooperation and meticulous attention to editing, design and layout. Our book would not have been possible without Jaime Plaza's expertise and enthusiasm in providing a major proportion of the photographs, and we are most appreciative of his vital contribution and willing help. We are grateful to Ron Oldfield for permission to use copies of his images of seeds and ants, Sue Bullock for the mycorrhizal cross section, Lyn McDougall for drawing the maps, Nicola Oram for the sandstone landscape cross section, and David Hardin for a number of photographs of particular species. Images by these people are supplemented with photographs by the authors. Our particular thanks go to Tim Entwisle, Director of the Plant Sciences Branch of the Royal Botanic Gardens, for his constructive comments and enthusiastic support, and other staff of the Royal Botanic Gardens who have provided information, advice and comments, especially Bob Coveny, Janelle Hatherly, John Lennis, Alistair Hay, Ken Hill, Peter Weston and Peter Wilson. We also thank most appreciatively Malcolm Reed and Barbara Rice for helpful comments and suggestions; Sue Gould, Peter Mitchell, Peter McGee, Charles Morris, Lesley Hughes and Mark Westoby for sharing results of their scientific research with us; Brian Atwell for his help; and staff of the NSW National Parks and Wildlife Service for providing information.

Publication Details

ISBN 0731393422

Published by the Royal Botanic Gardens Sydney 2000

Written by Jocelyn Howell & Doug Benson
Principal Photography by Jaime Plaza
Edited by Penny Farrant
Designed by Helen Stevenson

Printed by McPherson's Printing Group

Cover: Coastal sandstone heath with *Actinotus helianthi* Flannel Flowers on the Manly Scenic Walkway looking towards the Heads of Sydney Harbour. Photo: Jaime Plaza

Contents

A CLOSER LOOK AT
SYDNEY'S BUSHLAND

AT THE BEGINNING of the new Millenium we have unparalleled access to information. So much so that its acquisition has become a measure of status and power. At times it seems as though we have all the answers, and it is just a matter of finding the right book or website.

This book provides information but not all the answers. We see it as a springboard for discovering more about Sydney's diverse bushland. We hope it will encourage you to look more closely when you are visiting the bush and seeing the plants for yourselves. When you are experiencing the plants' habitats — coastal headlands, windswept hilltops, rocky hillsides, sheltered valleys — you may find yourself wondering — why? Why are some plants growing here on the hillside, while different ones grow along the creek? Why are some plants only on sandy soils, but others only on clay soils? Why do some grow in both sandy soils and clay soils?

As scientists, we don't know the answers to all the questions. But we encourage you to journey with us — not only across the landscape, but also into the fields of ecology and prehistory — to consider the forces and events that have shaped what we see today. Plant origins and diversity, plant links with other organisms, plant ways of surviving fires and floods — these are the interwoven themes that make up the rich tapestry of natural vegetation.

Our book is divided into three parts. In the first, we introduce Sydney's landscape and different types of bushland. The second focusses on aspects of bushland ecology — essential elements of the rich tapestry. Lastly, our book takes you out and about to places where bushland and its diversity can be experienced.

Take a closer look at Sydney's bushland — discover there is more than meets the eye!

◀ Patterns of heath and woodland form a variety of bushland habitats on the sandstone plateau landscape of Ku-ring-gai Chase National park in Sydney's north. Photo: Jaime Plaza

Geology shapes our landscapes and soils

Back in the Triassic when the delta was fantastic

Back in the Triassic period, around 200 million years ago, geological events set Sydney's shape in stone, literally! The future Sydney was then part of a massive river delta, with huge amounts of sand being deposited by repeated floods. After the sand, large amounts of clay were thinly layered as the delta silted up. Over time and with the weight of accumulated deposits on top, the sand was compressed into sandstone of the Hawkesbury and Narrabeen formations, and the clay into layers of shale — the Wianamatta Series — covering it.

The massive sandstone, up to 250 metres thick, was resistant to volcanic intrusions and to movements in the earth's crustal plates. However, when these 'tectonic' movements were very strong as Australia separated from the rest of the supercontinent Gondwana over millions of years, a line of weakness developed in the sandstone, and the part of it now under western Sydney gradually sank. Today we know this area as the Cumberland Plain. An abrupt change along the line of weakness, now known as the Lapstone Monocline, forms the eastern edge of the extensive sandstone plateau of the Blue Mountains. North and south of the Cumberland Plain, sandstone with its shale covering was left sloping gradually up to elevated plateau surfaces that then began to erode.

The Cumberland Plain was protected from erosion by its lower elevation, and so today is still covered by thick shale. However, erosion has removed the soft shale covering from all but the most central parts of the surrounding

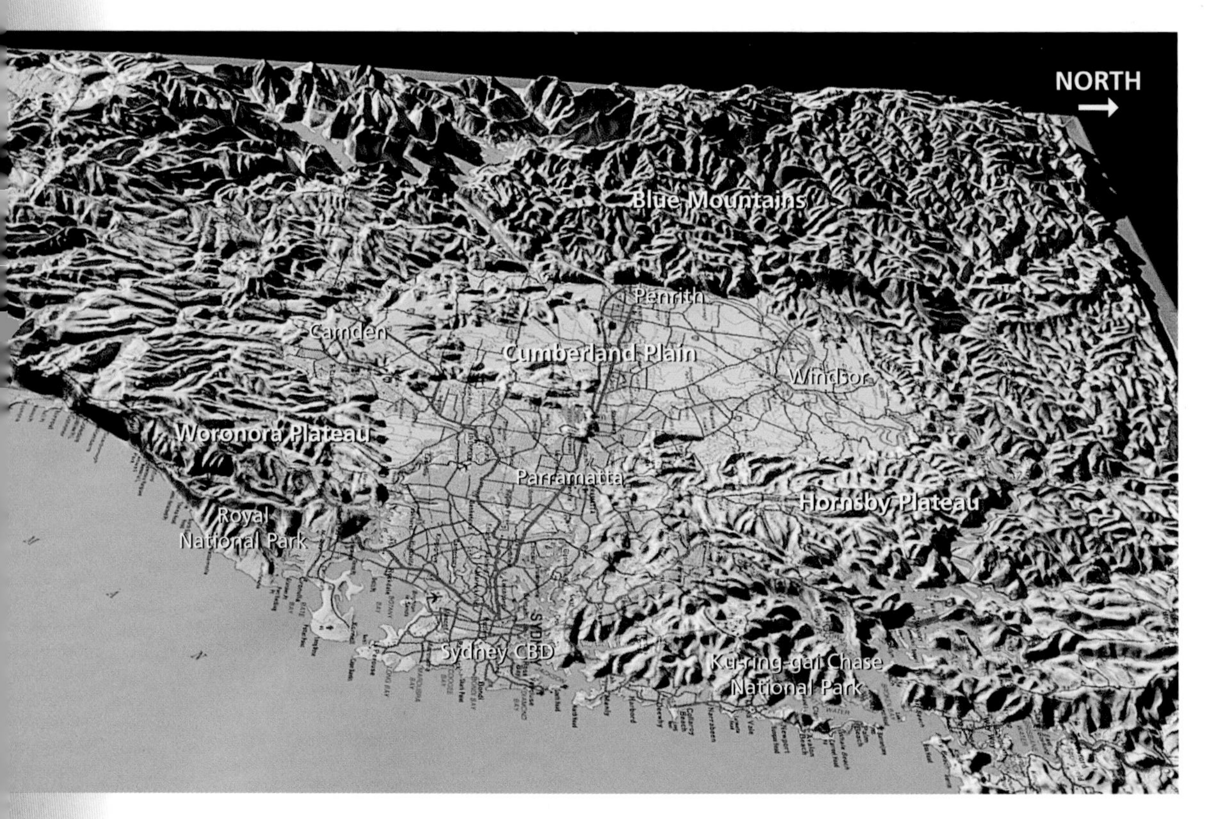

▲ Shale and sandstone landscapes can be clearly distinguished in this **relief map** of the Sydney region — the lumpy bits around the edges are sandstone, the smooth centre is shale.

elevated plateaus, leaving a landscape pattern of level shale-capped ridgetops flanked by rugged sandstone hillsides.

◀ The different features of **shale and sandstone** can be seen exposed in road and railway cuttings. The shale, at the top, is composed of compressed clay in thin layers, and is softer and more easily eroded. It weathers to clay-rich soil of fine texture and reasonable fertility. The sandstone beneath is composed of thicker strata of compressed sand, and is hard and resistant to erosion — notice the channels for explosives needed to make this cutting. The sandstone forms coarse-grained sandy soil, of very low fertility.

▲ In cuttings and at active erosion sites, such as here at The Gap, near South Head, you can see horizontal bands of sandstone indicating sand layers deposited by repeated ancestral floods. Vertical lines mark weaknesses where water has penetrated and contributed to erosion. If you look closely at cuttings or rock outcrops, you may see sloping lines within some of the bands. These lines are called crossbedding, and they mark sand deposited by periodic strong current flow. Sand layers deposited by turbulent water flow show less internal structure, and often weather to beautiful 'honeycomb' patterns.

Sandstone and shale set Sydney's scenery

Most of Sydney sits on sandstone and shale. These sedimentary rock types give rise to very different landforms and soils, each providing very different growing conditions for plants. Sandstone and shale are responsible for the distribution patterns of most of Sydney's native vegetation. The characteristics of these different landscapes and their soils also largely determined the historical patterns of settlement. Today's bushland remnants are the survivors of those events.

▲ The typical rugged **sandstone landscape** consists mainly of steep hillslopes between level ridgetops and rocky creek gullies, as here at Glenbrook Creek in the Blue Mountains. Shallow pockets of soil amongst boulders and outcropping rock are no use to farmers, so much bushland on sandstone landscapes has survived.

▲ In contrast, the characteristic **shale landscape** has gently undulating hills and is mostly cleared. For example, look at this view from Mt Annan, near Campbelltown. The first 'Europeans' to settle this fertile district were escapee cattle from early Sydney town. Colonists found the cows in the 1790s and named the place The Cowpastures.

The shale landscapes with their gentle slopes provide fairly uniform growing conditions for plants. There are differences between habitats on hillsides and along creeklines, for instance, but the greatest variation across the landscape is due to the decreasing rainfall gradient from over 1200 mm on the coast to as low as 700 mm in western Sydney.

Shale landscapes form on Wianamatta Shale, while the sandstone landscapes are formed from Hawkesbury Sandstone and the underlying Narrabeen Group sandstones. The Narrabeen Group sandstones also include shale-rich strata, and may outcrop in deep valleys towards Sydney's edges, for example at Bola Creek in Royal National Park, on lower hillslopes of Sydney's nothern beaches, along the lower Hawkesbury River, in the Gosford area and in the Grose Valley in the Blue Mountains. In these situations, the combination of clay-rich soil and shelter leads to good conditions for plant growth.

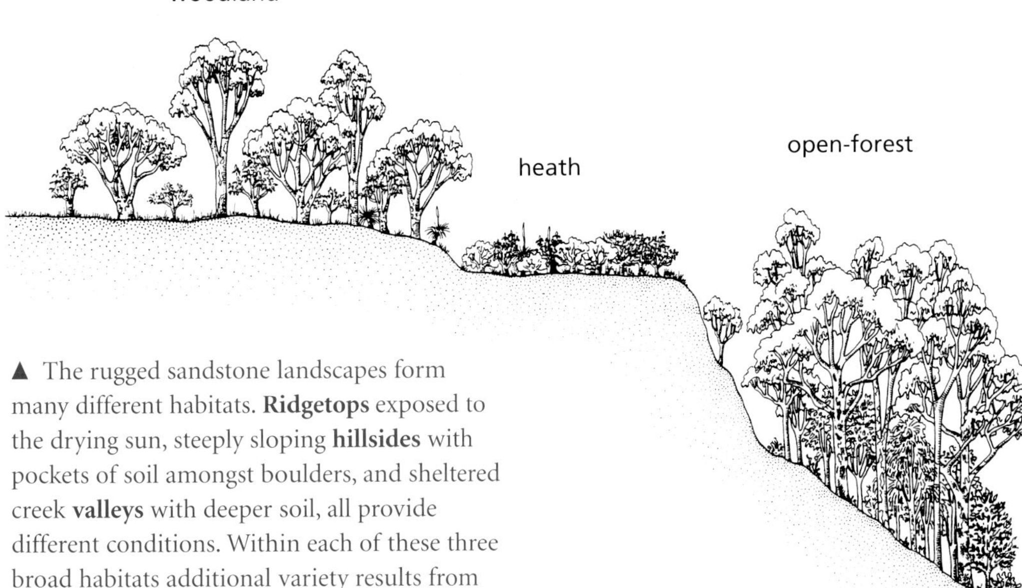

▲ The rugged sandstone landscapes form many different habitats. **Ridgetops** exposed to the drying sun, steeply sloping **hillsides** with pockets of soil amongst boulders, and sheltered creek **valleys** with deeper soil, all provide different conditions. Within each of these three broad habitats additional variety results from different aspects and soil depths. In addition, shale lenses within the sandstone strata add further local soil variation.

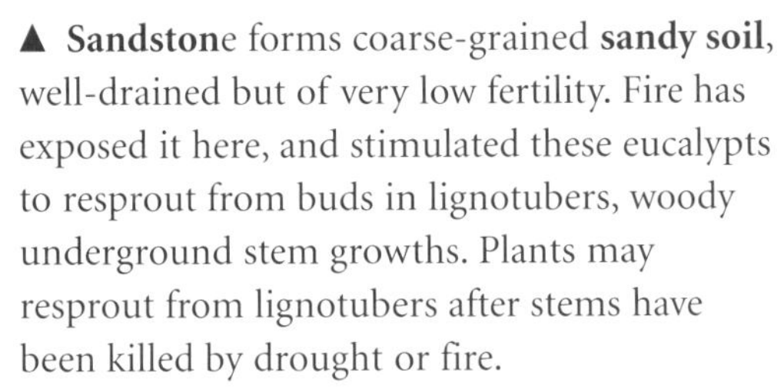

▲ **Sandstone** forms coarse-grained **sandy soil**, well-drained but of very low fertility. Fire has exposed it here, and stimulated these eucalypts to resprout from buds in lignotubers, woody underground stem growths. Plants may resprout from lignotubers after stems have been killed by drought or fire.

▲ **Shale** weathers to **clay-rich soil** of fine texture and reasonable fertility. The soil swells and shrinks with changes in moisture levels, and in western Sydney may be dry for long periods. Many herbs growing here die back during dry times and resprout from bulbs, tubers or thickened roots after rain — for example, the mauve-flowered *Brunoniella australis* next to rabbit diggings here.

Water and wind reshape the sediments

Landscapes formed by rivers

There are other landscapes in Sydney besides those formed directly by sandstone and shale.

Along the rivers and creeks are the alluvial landscapes, formed from alluvium — sand, silt, clay and gravel deposited by rivers during floods. Alluvium forms terraces and floodplains, and, except for the actual sloping banks of active rivers and creeks, these landscapes are generally flat.

▲ **Floodplains** can be quite extensive where alluvium has been laid over the flat shale landscape of the Cumberland Plain, as can be seen in this view towards Penrith from the edge of the Blue Mountains. In contrast, there's not much space for floodplains to form beside steep hillsides of a sandstone landscape.

Alluvial landscapes are prominent along the Hawkesbury-Nepean River and the Georges River. Soils form from the alluvial materials deposited in successive layers on riverbanks and floodplains. Sydney's alluvial landscapes are of different geological ages, and the soils formed from them have different qualities.

Alluvium deposited by floods during the last 10 000 years (the Recent geological epoch) is known as **Recent alluvium.** In a typical landscape pattern there are levee banks adjacent to the river — from here the floodplain slopes gently away. At the lowest point, usually furthest from the river, there may be a swamp. There is variation in these landscapes, maintained because material is still being laid down.

▲ **Soils from Recent alluvium** are fertile, with relatively high nutrient content and good drainage characteristics. They are valuable for farming, so most were cleared of their native vegetation in the early days of Sydney's European settlement. These soils are also highly prized for landscaping and gardening. Nowadays, considerable quantities are dug out and trucked away from riverbanks — like these at Menangle — for use as topsoil or building material.

◄ Older **Tertiary alluvium** can be seen in this exposed riverbank, with different proportions of gravel, sand and clay in the layers laid down by different floods. Soils formed vary in texture depending on which materials are exposed at the surface.

Tertiary alluvium, laid down during the Tertiary geological period, 2–65 million years ago, has been leached by rain and floods over millions of years. These weathered landscapes have very infertile soils, and are almost flat, with slight slopes to very shallow drainage lines.

In low-lying areas, wetlands form on alluvial soils. Freshwater wetlands form on river floodplains, while estuarine wetlands form where rivers are tidal in Broken Bay, Sydney Harbour, Botany Bay and Port Hacking.

Landscapes of windblown sand

Extensive sand sheets underlie today's south-eastern suburbs, stretching from Centennial Park to Botany Bay, and south of Botany Bay, from Kurnell to Cronulla. This is an aeolian landscape. Wind is believed to have deposited and shaped the sand, forming a dunefield much more extensive than similar areas that usually form behind beaches. Dunes and swales provide local habitat variety. This landscape had its own distinctive vegetation before it was taken over by Sydney's suburbs.

▲ North of Botany Bay most of the **aeolian landscape** has been obliterated by housing. However, some still survives here on the southern side of the bay in Botany Bay National Park. Organic matter accumulates in the swales, where freshwater soaks and swamps form, with wetland species adding to the richness of the sand sheet flora.

▲ At the surface the sand appears loose and well drained, but beneath the surface this aged sand has a structure. Water has leached minerals and organic matter down through it over the years to form a 'hard pan', a cemented layer a metre or two below the surface. You can see the top of the hard pan exposed as a yellowish and level surface in the foreground of this photo, where the white leached sand that normally lies above it has been removed. The leached sand and hard pan result in growing conditions different from those in sand dunes behind beaches or on sand directly above sandstone rock.

Volcanic landscapes ...

Landscapes formed from volcanic rock are uncommon in the Sydney area — the thick sandstone layer formed a barrier to volcanic eruptions, and generally only small intrusions of volcanic material made it through as diatremes, dykes or basalt flows. Small localised diatremes or dykes generally appear as holes or depressions in the landscape, while more extensive flows of lava over the sandstone have formed basalt caps, such as at Mt Wilson and Mt Tomah in the upper Blue Mountains. Because of the locally better soils, particularly where basalt occurs, most of these volcanic sites have been cleared of their original native vegetation. Many have also been quarried for construction gravel.

Climate is the other major influence on vegetation

Sydney is part of the subtropical east coast of Australia, and experiences a warm wet summer-autumn and a cool drier winter-spring. Local conditions vary according to landform features and distance from the sea. From the coast to the Cumberland Plain, rainfall follows a decreasing gradient, temperature extremes become more pronounced, and there is an increasing incidence of frost.

Rainfall is highest on the coast, where the annual average is over 1200 mm, and on the nearby elevated plateaus — with a maximum of 1440 mm per annum at Pymble-Turramurra. At Parramatta on the eastern margin of the Cumberland Plain, average annual rainfall drops to around 900 mm, while across the Plain's low-lying central basin, between Windsor and Picton, it is less than 800 mm. Between Campbelltown and Camden the annual average drops below 700 mm. At the eastern edge of the Blue Mountains, rainfall begins to rise again with altitude, reaching just over 1400 mm per annum in the upper Blue Mountains.

In seasonal terms, the Cumberland Plain experiences its wettest period during summer, while near the coast, the most rain generally falls in autumn. Rainfall is lower in winter, and spring is generally the driest season. Thunderstorms and hail, which may affect individual plant survival, have more influence on success of agricultural crops than on native plant distributions. There are about thirty thunderstorms per year, on average, in Sydney, most in late spring and summer.

Local **temperature** depends on aspect, altitude and distance from the coast. January is the hottest month and July the coldest. Mean maximum temperatures in January increase from less than 26°C along the coast to over 29°C on the Cumberland Plain and decrease to 23°C with increasing elevation in the Blue Mountains. Mean minimum temperatures for July drop from 7–8°C at Sydney to 2–4°C on the Cumberland Plain and Blue Mountains.

Frosts are rare on the coast but common further inland and may influence the distribution of some plant species. The duration of the frost period increases with increasing distance from the coast and, to a lesser extent, with elevation. On the Cumberland Plain the average frost period occurs between May and September and may exceed 100 days. In the Blue Mountains snow falls most frequently in July and August on about 3–10 days per year.

Variations in topography and distance from the sea produce marked variations in **wind** speed and direction. Main wind directions also vary with the season and time of day. For example, near the Sydney CBD in the afternoons, northeasterly or east-northeasterly winds are most common in all seasons except winter, when west-southwesterly winds are most frequent. However, westerly winds are likely to be the strongest.

Changing Climates

It is fascinating to speculate on how plant distributions we see today have been shaped by past climates as well as present. For example, at the peak of the last 'ice age' 18 000 years ago, climate was drier and colder, and sea level was lower, so that the coastline was further east. Mountain peaks and plateaus may have been covered in snow for extended periods, and only frost-hardy plants would have survived over large areas. Variations in climate have occurred over different time scales — tens of thousands of years, hundreds of thousands of years, millions of years.

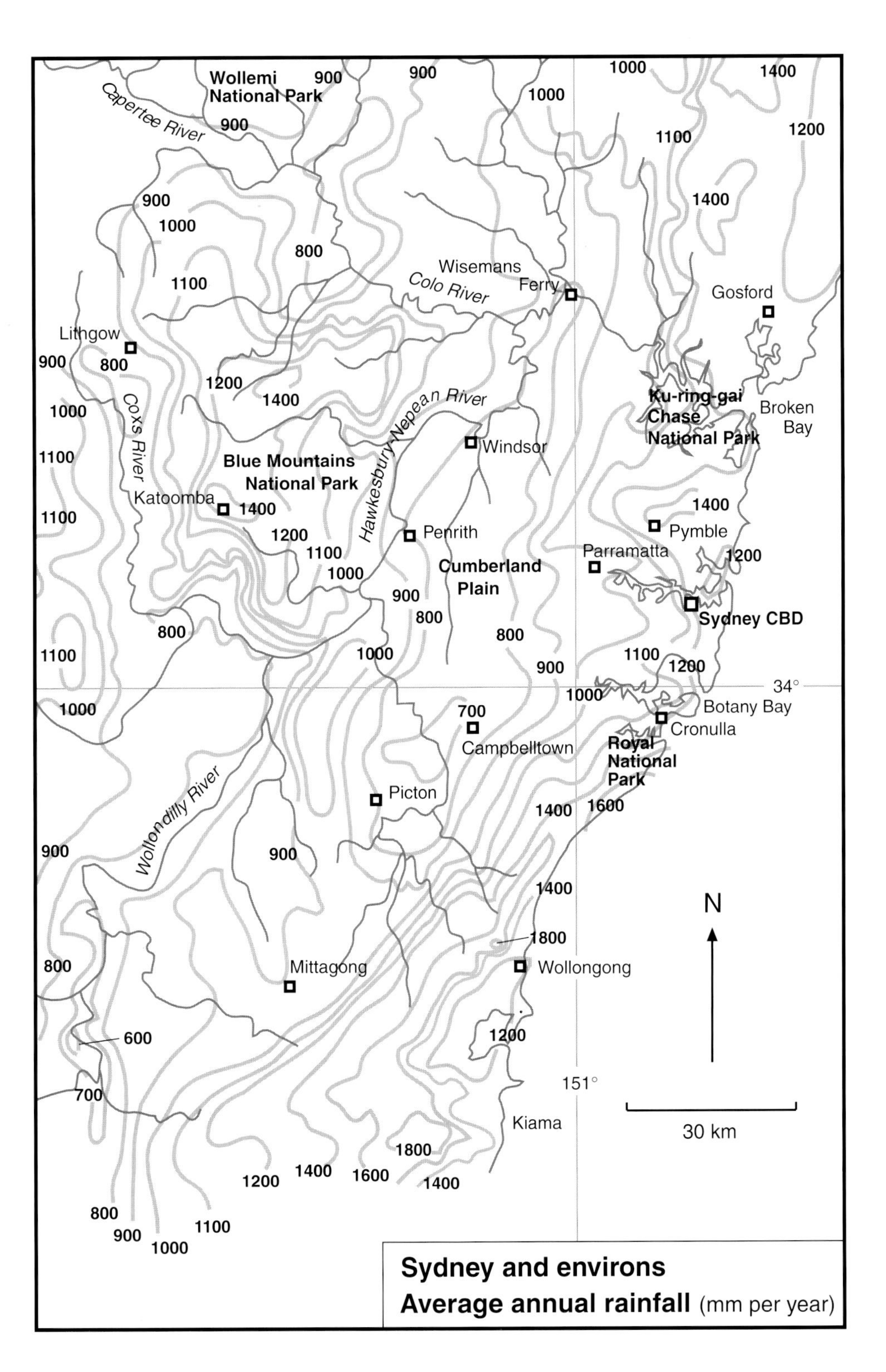

▲ Average annual rainfall (mm per year)

How do we describe the different types of bushland?

The height of the trees, how dense they are, whether there are trees at all — these features vary and we can use them to identify different types of vegetation. Forests and woodlands have trees, close together in forests but more widely spaced in woodlands. Heaths and scrub are made up of shrubs and other low-growing plants. Wetlands may be dominated by grass-like sedges or herbaceous plants.

Different types of forest, woodland and heath can be distinguished by the species they contain.There are about 2500 native 'vascular' plant species in the Sydney region — that is, plants with leaves, stems and roots. In this book we describe plants and vegetation growing on dry — or occasionally wet — land.

There are, however, other types of 'plants' growing in other places. **Mosses** and closely-related **liverworts** are tiny soft plants, without proper stems and roots, that grow in moist places such as gullies and rainforest. **Algae** have many forms — they may consist of only one cell, strings of cells, or may be multicellular and quite large, like the seaweeds. Algae grow in freshwater streams, lakes, and in the ocean. **Fungi** are often thought of as plants, but have more similarities to animals — such as living on food made by other organisms. We recognise some of them by their fruiting bodies, mushrooms and toadstools. Most fungal growth is out of sight however, for example, under the soil or in rotting wood. **Lichens** grow on rock or wood surfaces — some look like crusty or flaky coloured patches, some look mossy. Each lichen is made up of an alga and a fungus growing together in a mutually beneficial way. Mosses, algae, fungi and lichens are important in our bushland, and contain such variety that they deserve books of their own.

Where are the different types of bushland?

The bushland that once covered the Sydney area can be classified into broad types. You can see these on the map inside the front cover. We have pieced together the picture prior to European settlement of Sydney in 1788, using a variety of historical clues together with information on today's bushland remnants. This map of bushland types is like a simplified map of Sydney's geology, because the distributions of plants and types of vegetation are closely tied to geology.

Superimposed on the geological landscape are the effects of rainfall and temperature. Differences in average rainfall exert their effect on species distribution in a number of ways — one of the more important is the length of time that a plant must survive between falls of rain. On the Cumberland Plain sequences of dry days are likely to be longer than on the coast.

Sydney's bushland plants today are the survivors of evolution and past climatic change. Impending global warming caused by human activities may change plant distributions yet again at a much more rapid rate than in the past. Ecological research on plant life cycles and tolerances and interactions with insects will give some basis for predicting possible changes, but many other factors are involved.

► **Mosses,** on the right, and **lichens,** grey patches, grow together with the **Rock Felt Fern** ***Pyrrosia rupestris*** on this rainforest boulder.

▲ Sydney's bushland provides a dramatic example of the effect of rainfall on vegetation. Close to the coast where rainfall is high, 1200–1450 mm per annum, **tall open-forest** known as Blue Gum High Forest grows on fertile clay-rich soils on shale. Rainfall is high enough for closely spaced trees to grow up to 40 metres tall, and soft-leaved shrubs and ferns grow in the understorey.

▲ Only 60 kilometres from the coast, average annual rainfall drops by half, and the vegetation reflects this difference. At Mount Annan Botanic Garden near Campbelltown, where there is also fertile clay-rich soil, rainfall of around 700 mm supports only **woodland** growth — this is Cumberland Plain Woodland, with shorter, widely spaced trees above an understorey of grasses and spiky shrubs.

A plant by any other name?

All native plant species have scientific names; many have common names as well.

The trouble with common names is — they are not always common. The same plant may be called, for example, 'Prickly Mud Thistle' in the east, and 'Spiny Swamp Daisy' in the west. People using these names may be talking about the same plant without knowing it. Similarly, different people may use the same name for different plant species — 'Shivery Grass', for example. Different people using this name may think they are talking about the same plant but they actually have different species in mind.

The good thing about scientific or botanical names for plants is that they specify just one species of plant. Trouble is, many people find scientific names intimidating, and are afraid to try using them. Just think of them in the same way as names of people — of course it's hard to learn and remember a lot of new names all at once. But if you try learning just a few at a time, you will be surprised at just how quickly they become friends rather than obstacles. Perhaps you have trouble pronouncing scientific names. Don't worry! There is no recognised correct pronunciation — pronounce them as best you can and have a laugh with others over the different pronunciations. Scientific names are words in Latin, but they may be derived from Latin, Greek, place names, or the names of famous botanists and their friends.

By convention in writing, scientific names are usually in italics. The genus always comes first, with a capital, the species name is second, all in lower case, for example *Telopea speciosissima* and *Imperata cylindrica*. The species name is like the given name of a person, and denotes a single entity. The genus (plural: genera) name is like a person's family name, and many species may belong to a single genus. Related genera are grouped together into families. Family names end in '-aceae' as in Myrtaceae, the eucalypt or myrtle family.

Let's look at some plant communities, beginning with ...

Rainforest

Rainforest grows where there is fertile soil and plenty of moisture, and so is not very common around Sydney. We find it only in places such as volcanic outcrops, and in moist sheltered gullies on shale soils and alluvium. In rainforest, trees grow close together and their canopies shade out 70% or more of the sky — only dappled sunlight can penetrate. Technically rainforest is called closed-forest.

▲ **Rainforest** trees, like those lining the creek here, have dense canopies of soft, dark green leaves. In Sydney rainforest, the most common large trees are Coachwood *Ceratopetalum apetalum*, Sassafras *Doryphora sassafras* and Lilly Pilly *Acmena smithii*. In the sandstone landscape, rainforest species grow in narrow bands along creeks where there is permanent moisture and nutrient-enriched soil.

◀ Like many rainforest plants, the fruits of the Lilly Pilly ***Acmena smithii***, have fleshy coats and are dispersed by birds and other animals. It is one of a group of fleshy-fruited plants in the Myrtaceae family, which also includes the ecualypts.

▲ **Lichens** may form beautiful patterns on the tree trunks of Coachwood ***Ceratopetalum apetalum.***

◀ The small tree ***Synoum glandulosum*** has 3-lobed capsules containing shiny bronze seeds with bright orange arils, attached food bodies that encourage birds and other animals to disperse the seeds. Its pinnate leaves are characteristic of the family Meliaceae, which also includes the larger rainforest tree, Red Cedar *Toona ciliata*, over-exploited for its beautiful timber, and the White Cedar *Melia azedarach*, found mostly in dry rainforest and alluvial forest. *Melia* and *Toona* are unusual for Australian trees, as they are partly deciduous and lose their leaves before flowering.

At the rainforest floor there are only low levels of light. Treeferns flourish in the shelter beneath dense canopies. Vines and epiphytes — plants that grow on other plants — are characteristic rainforest plant growth forms, resulting from the competition for light. Thick stems of vines coil upwards into the tree canopies.

▲ In mountain rainforest, as here at Mount Tomah, **ferns** such as *Microsorum scandens* carpet the forest floor, and some grow up tree trunks as epiphytes.

◄ Soft Treefern ***Dicksonia antarctica*** is widespread in mountain rainforest gullies. Its unfurling fronds form beautiful patterns.

► Orange Blossom Orchid ***Sarcochilus falcatus*** is one of several epiphytic orchids that grow on rainforest tree trunks.

► Palms are another feature of rainforest. The Cabbage Palm ***Livistona australis*** occurs in rainforest and swamp forest. The other Sydney palm, Bangalow Palm *Archontophoenix cunninghamiana* is restricted to very sheltered spots on the coast.

Rainforest fascinates Australians because it contrasts with our generally dry harsh environment. It is also perhaps the least characteristically Australian type of vegetation, because of the absence of the familiar gum trees and the presence of jungle or wet tropical species of trees, vines, ferns, orchids and epiphytes. Because of this, rainforest was once thought to have invaded Australia from the north but is now recognised as being an ancient part of the flora and to have coexisted with the sclerophylls — the eucalypts and other hard-leaved plants — since the breakup of Gondwana.

The lands of the gum tree — the open-forests and woodlands

In contrast to the dark green closed canopies of the rainforest, much of Sydney's forest is open-forest dominated by the open canopies of the ubiquitous gumtree or eucalypt. Eucalypts are plants of the closely related genera *Eucalyptus*, *Angophora* and *Corymbia* in the Myrtaceae or myrtle family. These species are mainly trees, and in open-forest their canopies touch or overlap, shading out 30–70% of the sun.

Moist tall open-forest grows on high nutrient soils where there is high rainfall, but where conditions are not favourable enough for rainforest. Soils on shales, fertile alluvium, and basalt provide the nutrients necessary. Traditionally, moist tall open-forests have been known as Wet Sclerophyll Forest and drier open-forests as Dry Sclerophyll Forest.

◄ One of the understorey trees is Forest Oak, ***Allocasuarina torulosa***. It has cork-like bark, purple-tinged new growth, and often has long drooping young branches. Woody capsules open after fire to shed their winged seeds. Without fire, capsules eventually drop from the tree and open to release seed. Seedlings may establish in the absence of disturbance.

The Blue Gum High Forest of Sydney's north shore

Blue Gum High Forest once covered Sydney's north shore plateaus where rainfall is highest and deep shale covers the sandstone. Only a few small remnants have survived Sydney's development, and Blue Gum High Forest is now listed as an Endangered Ecological Community under the *NSW Threatened Species Conservation Act*.

▲ The tallest trees in Blue Gum High Forest grow up to 40 metres high. Sydney Blue Gum ***Eucalyptus saligna*** has smooth bark except for a 'sock' of rough bark on its lower trunk.

▲ Blackbutt ***Eucalyptus pilularis*** is the other dominant tree of the Blue Gum High Forest. The rough bark on its trunk is finely fibrous, in contrast to its smooth branches. Like Sydney Blue Gum, the Blackbutt was used as construction timber in Sydney's early colonial days.

◀ Ferns growing in the moister parts of the forest include Gristle Fern, ***Blechnum cartilagineum***. Aboriginal people are believed to have roasted rhizomes — creeping stems — of this fern for winter food.

◀ Red-fruited ***Stephania japonica*** vine twines around Bracken Fern *Pteridium esculentum* growing in drier parts of the forest.

▲ Stiff leaves of the scrambling pea ***Platylobium formosum*** contrast with soft leaves of many other plants in the Blue Gum High Forest.

◀ Flowers of the native Ivy-leaved Violet ***Viola hederacea*** brighten the forest floor in spring and summer.

▲ ***Morinda jasminoides*** may begin life as a straggly shrub before developing into a woody climber. Its fragrant white flowers and opposite leaves are characteristic of the family Rubiaceae. Lumps on the leaves — called 'domatia' — are homes to tiny spider-like mites.

▲ The ground orchid ***Acianthus fornicatus*** sends up flowering spikes in winter. Its heart-shaped leaves, inconspicuous in the forest floor litter, and its pale brown flowering stalks are a challenge for the observant bushwalker.

Very heavy timber, chiefly iron and stringybark ...

Sydney Turpentine-Ironbark Forest

Sydney Turpentine-Ironbark Forest is not as tall as Blue Gum High Forest. It is open-forest that grows where shale covers the sandstone more thinly and/or rainfall is lower. It once covered much of the inner south-western suburbs between the coast and Parramatta, and the shallower shale plateau areas to the north and west of the North Shore. Most was cleared for farming and suburbs, and it is now listed as an Endangered Ecological Community under the *NSW Threatened Species Conservation Act*. The dominant trees are Turpentines ***Syncarpia glomulifera*** and Grey Ironbarks ***Eucalyptus paniculata***, which grow up to 30 metres tall.

▲ Here a Grey Ironbark (trunk on the left), and a Turpentine (broad leaves on the right), grow above *Pittosporum undulatum*, in **Sydney Turpentine-Ironbark Forest**.

▲ Broad oppositely-arranged leaves and 'gun-turret' fruits distinguish the Turpentine, ***Syncarpia glomulifera***, from eucalypts. It has thick stringy bark that is soft to the touch.

◄ In contrast to the fibrous bark of the Turpentine, that of the Grey Ironbark, ***Eucalyptus paniculata***, is firm, deeply furrowed, and hard to break off. This long-lived tree grows up to 30 metres tall. Its tough timber was used by settlers for construction work and fencing.

◄ ***Pittosporum undulatum,*** Sweet Pittosporum, is a small native understorey tree of Sydney Turpentine-Ironbark Forest. It is common in the changed conditions along the edges of urban bushland, as it germinates in the absence of fire and grows quickly in moist, low light conditions.

▲ Different flowering times help distinguish two similar understorey wattle trees — *Acacia decurrens* which flowers in August, and ***Acacia parramattensis***, pictured, which flowers in December.

▲ The small tree ***Clerodendrum tomentosum***, in the Verbena family Verbenaceae, has fragrant white flowers in November. They are followed by purplish-black fruits in colourful red bracts.

▲ Bright red pea flowers give the vine ***Kennedia rubicunda*** one of its common names, Running Postman.

▲ Shrubby ***Pittosporum revolutum*** is slower-growing and less common than *Pittosporum undulatum*. Its orange fruits open to reveal bright red sticky seeds.

A tangle of vines and dense undergrowth

River-flat forests of fertile alluvial soils

River-flat forests grew on fertile soils where riverbanks and floodplains were formed from Recent alluvium, along the Hawkesbury-Nepean and Georges Rivers. These deep, well-drained soils were discovered by Sydney's early colonists and rapidly cleared.

Sydney Coastal River-flat Forest is listed as an Endangered Ecological Community under the *NSW Threatened Species Conservation Act*. A few small forest remnants survive, for example, in Belgenny Reserve at Camden and in Mitchell Park (in Cattai National Park) on the Hawkesbury-Nepean River system, and at Deepwater Park, Milperra on the Georges River.

▲ Towering above understorey *Acacia parramattensis* and *Tristaniopsis laurina* trees in River-flat forest near Camden are the distinctive twisted branches of ***Angophora subvelutina*** Broad-leaved Apple. It grows sporadically on fertile levee banks.

▲ ***Eucalyptus baueriana*** grows from Menangle to Wallacia on the Nepean, and around Milperra on the Georges River. Its common name Blue Box comes from the distinctive rounded blue-green leaves.

◄ River-flat forest trees grow to over 30 metres tall, like this veteran Blue Gum, ***Eucalyptus botryoides/ saligna*** at Camden Park, which may be over 200 years old. Blue Gum populations south of Sydney show genetic influence from *Eucalyptus botryoides.*

Eucalyptus amplifolia, Eucalyptus deanei, Eucalyptus elata, Eucalyptus tereticornis and *Eucalyptus viminalis* are other river-flat forest trees.

▲ Willow Bottlebrush ***Callistemon salignus*** is a 'willow-leaved' riverbank plant with cream flowers and pink new leaf growth. It is widespread but not very common.

▲ In the forest understorey, mauve flowers appear in September-October on the White Cedar ***Melia azedarach***, followed by honey-coloured fruits. This is one of Sydney's few deciduous native trees.

▲ Attractive fruits of velvety-leaved shrubby small tree ***Clerodendrum tomentosum*** enliven the understorey in January-February.

◄ A prickly-stemmed small tree, ***Hibiscus heterophyllus*** grows well in places like riverbanks that are likely to be disturbed. Aboriginal people used stem fibres for making string.

◄ ***Smilax australis*** is known as Lawyer Vine because its tenacious spines don't readily let you go. It is one of the understorey vines that helped make walking on the riverbanks 'very fatiguing' for early European explorers.

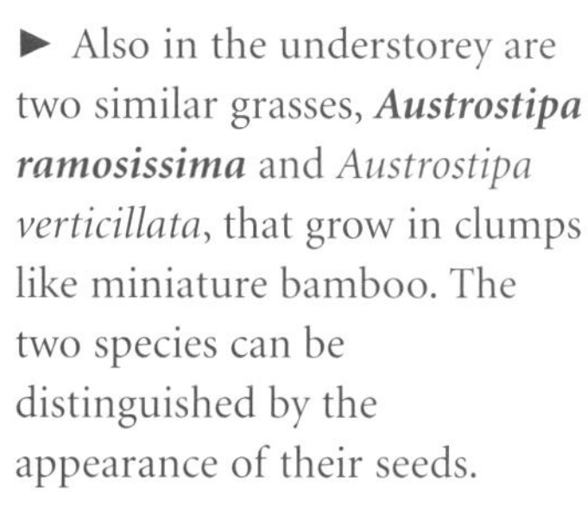

► Also in the understorey are two similar grasses, ***Austrostipa ramosissima*** and *Austrostipa verticillata*, that grow in clumps like miniature bamboo. The two species can be distinguished by the appearance of their seeds.

► River Oaks ***Casuarina cunninghamiana*** are the main trees along the water's edge, and may achieve impressive stature. These trees now in Cattai National Park were probably planted about 50 years ago, at a time when native trees were rarely planted along the riverbanks.

By channels of coolness ...

Tall open-forest in sheltered sandstone gullies

Tall open-forest also grows in sheltered sandstone gullies, where soils have been enriched with nutrients washed down from shale on the ridgetops. Tree species depend on how favourable conditions are in these gullies. This means that some species grow in several different types of tall open-forest.

▲ Here, for example, is Blackbutt ***Eucalyptus pilularis***, growing in good soil and moist conditions in a deep sheltered sandstone valley. Blackbutts also grow in the fertile clay soils on shale-capped ridgetops.

Other trees that may grow where soil is fertile in sheltered sandstone valleys include Turpentines *Syncarpia glomulifera,* Blue-leaved Stringybarks *Eucalyptus agglomerata* and Sydney Blue Gums *Eucalyptus saligna.*

▲ Deane's Gum ***Eucalyptus deanei*** grows in fertile soil at the base of sandstone hillsides along the Hawkesbury-Nepean River and tributaries, as well as on shale cappings, and volcanic necks in the lower Blue Mountains. It is closely related to Sydney Blue Gum *Eucalyptus saligna.*

▲ Common in moist forests is Giant Water Vine ***Cissus hypoglauca.*** It has distinctive 5-part leaves. It belongs in the grape family, Vitaceae, and develops bunches of purplish-black fleshy fruits.

▲ Wonga Wonga Vine ***Pandorea pandorana*** grows around Sydney in Blue Gum High Forest, Sydney Turpentine-Ironbark Forest, River-flat Forests, coastal rainforest and moist open-forest of sandstone gullies. It is also propagated and sold in nurseries as a popular garden plant.

▲ The vine Traveller's Joy ***Clematis aristata*** has white starry flowers. The seeds form in clusters, each with a feathery plume — hence its other common name of Old Man's Beard.

▲ ***Calochlaena dubia*** resembles Bracken Fern *Pteridium esculentum*, but has softer, paler green fronds and grows in moister, more sheltered sites. Like many ferns, *Calochlaena* spreads by means of rhizomes — horizontal stems — from which the fronds rise, to form dense fern banks in sheltered open-forest.

The transition between moist tall open-forest and rainforest may be sharp or gradual depending on the site and on the effects of fire. Tall open-forests are able to cope with occasional fires, and may depend on them to provide the open conditions necessary for eucalypt seedlings to establish. On the other hand, rainforest is set back by fire.

Dimpled trunks and twisted branches

Open-forest of the sandstone hillsides

Lower-growing open-forest occurs extensively in the sandstone landscapes. Indeed, **open-forest of the sandstone hillsides** is the type of forest most commonly encountered around Sydney, as there is little incentive to clear the rugged sandstone hillsides. This forest grows where conditions are intermediate between the woodland of the upper, exposed slopes and the tall open-forest of the moist, sheltered gullies. Open-forest varies in composition depending on the aspect. In Australia, north and west facing slopes are sunnier and drier than those facing east and south. More shaded slopes provide sheltered, moister conditions for plants.

▲ Smooth pink dimpled trunks belong to ***Angophora costata***, Smooth-barked Apple or Sydney Red Gum, the most instantly recognisable tree of this open-forest. Its characteristic canopy of twisted branches help make it one of the best-loved and most-photographed trees in Sydney's bushland.

◄ Less flamboyant but a constant companion in hillslope open-forest is the Sydney Peppermint ***Eucalyptus piperita***. You may often see ribbons of peeled bark hanging from its smooth-barked, straighter upper branches. The finely-fibrous bark on trunk and lower branches contrasts with the flaky, tessellated bark of the Red Bloodwood, ***Corymbia gummifera*** on the left.

◄ In more sheltered hillside places, delicate white fringed flowers of Blueberry Ash, ***Elaeocarpus reticulatus***, appear in November, followed by attractive blue fruits.

◄ Around Christmas time, Christmas Bush, ***Ceratopetalum gummiferum***, takes on orange-red tints that match the peeling trunks of *Angophora costata*. The colour belongs to petal-like bracts that enlarge and redden after flowering.

As Christmas Bush sheds its colourful bracts, the bright orange of *Angophora costata* trunks begins to fade to delicate shades of pink and grey.

▲ Among the ***Angophora costata* trunks** is a variety of understorey shrubs. In August and September the sandstone hillsides are alive with a multicoloured wildflower display. In late summer, however, Sunshine Wattle ***Acacia terminalis*** is one of the few shrubs flowering.

▲ After the initial springtime wildflower exuberance, sheltered slopes may become suffused with the orange-gold of ***Pultenaea flexilis*** pea-flowers in October.

◄ The small yellow flowers of Pine-leaved Geebung ***Persoonia pinifolia*** develop into these chewy "geebungs", one of the few fleshy fruits on the sandstone hillsides. *Persoonia* is unusual amongst the local members of the Proteaceae family, most of which have dry or woody fruits.

▲ ***Allocasuarina littoralis***, the Black Sheoak, is a small tree of open-forest and woodland on sandstone, often forming dense thickets associated with the clay soil of deep shale lenses. Bands of shale occur sporadically through the Hawkesbury Sandstone and outcrop on ridgetops and hillsides. Dense layers of *Allocasuarina* 'needles' accumulate beneath the trees, and appear to physically inhibit seedlings of other plants.

Bent and twisted by floods

Closely associated with the open-forests of the sandstone hillsides and gullies is riverine or **riparian scrub**. It occurs in narrow bands amongst sandstone boulders along more-or-less permanently flowing creeks. The scrub has a distinctive group of species, including multi-stemmed shrubs that can survive bending and twisting caused by periodic flooding.

◄ The bright yellow summer flowers of ***Tristaniopsis laurina*** have made it popular as a street and garden tree.

▲ Possibly the most distinctive and ubiquitous plant in the **riparian scrub** is *Tristaniopsis laurina*, the Water Gum. Its mottled smooth trunks are often twisted and bent over amongst the boulders as a result of floods.

Among the most widespread species of this moist sandy creekside habitat are *Lomatia myricoides* River Lomatia and *Stenocarpus salignus* Scrub Beefwood in the Proteaceae family, *Acacia floribunda* in the Fabaceae and ground covers *Lomandra longifolia* and *Lomandra fluviatilis*.

◄ The small shrubby Black Wattle ***Callicoma serratifolia*** has round yellow flowers like an *Acacia*, but is in the Cunoniaceae family and so is related to Christmas Bush. Its serrated leaves — hence the name *serratifolia* — are particularly attractive when young and bronzed.

▲ Fronds of ***Sticherus flabellatus***, Umbrella Fern, form extensive 'carpets' in the shelter of trees in sandstone gullies where there is deeper sand and a reliable moisture supply.

▲ The teatree ***Leptospermum polygalifolium*** is widespread, but particularly abundant close to creeks, where it often assumes a 'leggy' form in dense scrub. It is a shrub with thin, hard, finely-fissured bark and abundant white flowers on pendant branches in October-December.

▲ In the upper reaches of creeks, almost always beside water, you may see the drooping pink flowers of the scrambling shrub ***Bauera rubioides***, named after the Austrian artist Ferdinand Bauer, who made exquisite paintings of native plants after visiting Sydney in the early 1800s.

Riparian scrub may be found grading into open-forest of *Angophora costata* and *Eucalyptus piperita* at the base of sandstone hillsides. Where gully alluvium provides favourable conditions, riparian scrub grades into moist tall open-forest or into rainforest.

Much of the 'ordinary' vegetation you'll see around Sydney is woodland

In woodlands, trees are widely spaced, their canopies shading out up to 30% of the sunlight. Woodlands occur where conditions are not good enough for trees to grow as densely as they do in forest — either soil is not so fertile or there is not as much rain. In Sydney, woodland occurs on sandstone, where there is enough rainfall for forests but where the soil is infertile, and on shale where soil is relatively fertile but the rainfall is insufficient. Woodland on sandstone has twisted, often multi-trunked trees and an understorey dense with shrubs. Woodland on shale has taller, straighter trees above an understorey rich in grasses, with shrubby patches and seasonal appearances of daisies, lilies and orchids.

Because sandstone weathers to a rugged landscape with shallow infertile soil, there is much more bushland remaining on sandstone than on shale. If you're standing in woodland in Sydney, it's most likely to be sandstone woodland. Woodland also occurs on Tertiary alluvium, on soils of in-between fertility where rainfall is low, and there's an in-between amount of this woodland remaining.

Woodlands on sandstone soils

On the rugged Hawkesbury Sandstone landscapes that surround Sydney, woodland is the main type of vegetation. It grows on the ridgetops and upper hillslopes, particularly the exposed slopes facing north and west, where ample sunlight dries out the soil. It is structurally very variable and includes areas of woodland, open-woodland, low woodland and low open-woodland, depending on local aspect, soil, drainage and time since the last fire.

Whilst eucalypts form the canopy species in woodland and open-forest on sandstone, species of the Proteaceae family are common in the understorey.

▲ Scribbly Gums, ***Eucalyptus haemastoma***, stocky trees with thick white trunks, often with large burnt-out basal cavities, are unmistakeable features of the ridgetop sandstone woodland. Here pink-flowering ***Boronia ledifolia*** is prominent in the understorey.

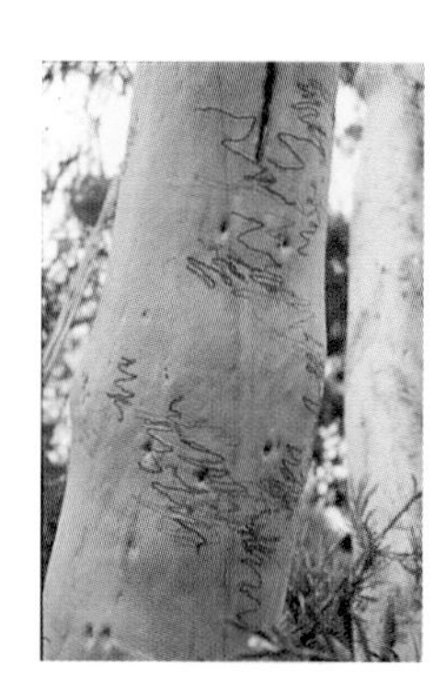

◄ Smooth-barked eucalypts are known as 'gums'. **Scribbly Gum bark** peels off each year to reveal a smooth cream surface covered in brown 'scribbles', chewed by moth larvae feeding just beneath the outer layer.

Where conditions are too harsh for normal tree growth, you may see mallee eucalypts, trees with many thin trunks arising from the ground instead of a single thick trunk. Some species may assume single-trunked or mallee form depending on conditions, while other species always grow as mallees. When mallee stems are killed by fire, new stems grow from the underground lignotuber or rootstock.

Species that always occur in mallee form on the Sydney coast include *Eucalyptus luehmanniana*, Yellow-top Ash, and *Eucalyptus obstans*, Port Jackson Mallee, while *Eucalyptus stricta*, Mallee Ash, is common in the Blue Mountains.

▲ ***Eucalyptus sieberi*** is known as Black Ash or Silver-top Ash because the thick black bark on its trunk contrasts with the smooth silvery branches. It can grow in either mallee or single-trunked form. Here on a rock ledge, with particularly thin soil, it is growing in mallee form.

▲ Bark helps to identify different eucalypts. Here flaky yellowish bark of the Yellow Bloodwood ***Corymbia eximia*** distinguishes it from Grey Gum ***Eucalyptus punctata*** which has a smooth grey-brown bark that sheds to reveal orange blotches.

◄ The rough, tessellated nature of bloodwood bark can be seen on the trunk of this Red Bloodwood, ***Corymbia gummifera***. The bark is often blackened from bushfires or stained with red resin exuding from wounds.

▲ ***Banksia serrata*** is a distinctive small tree in the Proteaceae family. It has tough serrated leaves, lumpy rough bark on a chunky trunk and large seed cones usually at a variety of stages. Attractive patterns form along the flower spikes as individual flowers open. Watch out for 'big bad banksia men' lurking behind the flowers — the old seed cones with openings like eyes were given personalities in children's stories by May Gibbs.

In the woodlands on sandstone soils ...

The smaller shrubs — such variety!

In August–September it is largely the smaller shrubs that have made Sydney's spring wildflower displays famous. The Fabaceae or pea family and the Proteaceae are particularly well represented in woodland understoreys, along with the Epacridaceae or southern heath family, and Rutaceae or citrus family. Ground orchids, lilies and other grass-like plants add to the diversity.

▲ Plants in the Proteaceae family include the well-known banksias and grevilleas. Also in the Proteaceae is prickly-leaved ***Lambertia formosa***, known as Mountain Devil because of its woody fruits shaped like horned heads. There is one among leaves towards the lower right of the picture. Despite its name, this Mountain Devil is not ferocious.

▲ ***Grevillea speciosa***, the Red Spider Flower, grows north of Sydney Harbour. Its eye-catching flowers hang clear of the leaves. A similar species that you may see south of the Harbour, *Grevillea oleoides*, has less prominent flowers and longer leaves.

▲ Another shrub in the Proteaceae family is ***Lomatia silaifolia***. It sends up spikes of cream flowers from foliage reminiscent of the culinary herb parsley. The decorative flattened fruits that follow contain winged seeds.

▲ Many pea family shrubs have flowers in shades of yellow. Among the largest are those of ***Gompholobium grandiflorum***, the Large Wedge Pea. Its buds look like small black eggs.

▲ There are often highlights of darker colours that play a role in guiding pollinators to pea flowers, like the red-brown markings on ***Bossiaea scolopendria*** flowers. This straggly shrub has flattened, winged stems about 1cm wide, with leaves reduced to tiny scales.

▲ Large pale pink flowers make ***Eriostemon australasius*** one of the best-loved plants in the family Rutaceae. It grows particularly well near rock outcrops.

◄ ***Thysanotus tuberosus***, Common Fringe-lily, is amongst the more delicate of the lilies and lily-like plants. Beneath the soil tiny tubers form as storage organs on its roots.

◄ In August the Dotted Sun Orchid ***Thelymitra ixioides*** sends up spikes of bright blue flowers from underground tubers. Fruit capsules are full of tiny transparent seeds.

Grassy woodlands of western Sydney

Cumberland Plain Woodland on the shale soils

The native vegetation on the clay soils of the Wianamatta Shale landscape of western Sydney is eucalypt woodland, known as Cumberland Plain Woodland. In contrast to the shrubby woodland of the sandstone landscapes, the understorey is predominantly grassy and herbaceous. Much of this woodland has been cleared for housing and farming.

Trees forming the Cumberland Plain Woodland canopy are Grey Box *Eucalyptus moluccana*, Forest Red Gum *Eucalyptus tereticornis* and Narrow-leaved Ironbark *Eucalyptus crebra*.

► Grey Box ***Eucalyptus moluccana*** — growing here among white-flowering ***Bursaria spinosa*** shrubs — has fibrous grey 'box' bark on trunk, and smooth branches.

▲ The dark furrowed 'ironbark' of ***Eucalyptus crebra***, centre, contrasts with the paler trunks of the other woodland trees.

Woodland trees may be host to mistletoes, semi-parasitic native plants mostly in the Loranthaceae family. Mistletoes have leaves but no roots. Birds attach the sticky mistletoe seeds to branches, where they grow and draw nutrients from trees.

► Forest Red Gum ***Eucalyptus tereticornis*** has mostly smooth bark, banded in grey and white.

▲ ***Bursaria spinosa***, Blackthorn, is the most common shrub in the Cumberland Plain Woodland understorey. Most of western Sydney's woodland remnants have been used for grazing in the past, and the thorns on *Bursaria* have made it unpalatable to stock. In summer its white flowers infuse the woodland with a delightful fragrance, giving rise to another of its common names, Sweet Bursaria.

◄ The ground cover in Cumberland Plain Woodland is mostly grassy and herbaceous. ***Themeda australis***, Kangaroo Grass, is widespread. Its decorative seed heads may give parts of the woodland a bronze sheen from spring to autumn.

◄ Yellow flowers of the Bulbine Lily, ***Bulbine bulbosa.*** Flowers of small herbs, lilies and orchids may appear like jewels amongst the understorey grasses in spring. These plants often die back in the hot dry summer months.

▲ Although the herb ***Ajuga australis*** may flower year-round, you are most likely to see its deep blue flowers in spring. A member of the mint family Lamiaceae, it is related to northern hemisphere Bugles.

Because of the extensive agricultural and suburban development of the shale soils of western Sydney, most Cumberland Plain Woodland has been cleared. It is now listed as an Endangered Ecological Community under the *NSW Threatened Species Conservation Act.*

A surprising place for plant richness

Castlereagh Woodlands on Tertiary Alluvium

During the Tertiary geological period, 2 to 65 million years before the present, ancestors of Sydney's Hawkesbury-Nepean and Georges Rivers laid down sequential layers of sand, silt, clay and gravel on their floodplains, in different patterns depending on the size and speed of the floods. Over millions of years nutrients in these Tertiary alluvial sediments were leached out, leaving the soils with very low fertility for plants.

The **Castlereagh Woodlands** is the name given to the characteristic vegetation that grows on the old Tertiary sediments that range from clays through to gravels and sands. This vegetation includes ironbark forest on the gravelly clayey soils, scribbly gum woodland on the sandier soils, and swamp woodland in the shallow drainage lines. Castlereagh Woodlands occur extensively between Windsor and Penrith in western Sydney and between Liverpool and Holsworthy in south-western Sydney.

Castlereagh Woodlands are surprisingly rich in species. There are 'endemic' species that occur only here on the Tertiary alluvial soils, and species that occur also on the inland western slopes. The Castlereagh Woodland flora also has similarities with sandstone woodlands because of the low nutrient status of the soils. As with most other Sydney bushland, periodic fires are part of this woodland's ecology.

◀ Ironbarks are the main trees on the clayey gravelly soils, especially the Broad-leaved Ironbark, ***Eucalyptus fibrosa.*** These often form denser forest as well as woodland.

◀ There is generally a shrubby understorey beneath the ironbarks. Amongst the spring-flowering shrubs is ***Prostanthera scutellarioides***, in the mint family Lamiaceae.

◀ ***Pultenaea parviflora*** is a rare dwarf shrubby pea listed as endangered under the *NSW Threatened Species Conservation Act.* It grows only on clayey soils in Castlereagh Woodlands in the Penrith-Windsor district.

A different group of plants grows in the poorly drained soils of ephemeral watercourses. This vegetation, the **Castlereagh Swamp Woodland**, has been listed as an Endangered Ecological Community under the *NSW Threatened Species Conservation Act* because of its restricted occurrence and the threats of clearing.

▲ Here the two most common small trees are recovering from a recent fire. ***Eucalyptus parramattensis,*** on the right, is resprouting with shoots on trunk and branches, and on the left, the papery bark of ***Melaleuca decora*** is still blackened.

▲ On the more sandy soils is the **Castlereagh Scribbly Gum Woodland**. Here the white bark of Hard-leaved Scribbly Gums, *Eucalyptus sclerophylla*, contrasts with the fibrous bark of Narrow-leaved Apples, *Angophora bakeri*, often burnt black. On the low-nutrient sandy Tertiary alluvial soils, understorey shrubs approach those of the sandstone flora in their variety. Yellow flowers of ***Acacia elongata*** appear in September.

▲ Another yellow-flowered shrubby pea, ***Dillwynia tenuifolia*** is virtually restricted to the Castlereagh Woodlands.

▲ The sturdy upright flower spikes of ***Banksia spinulosa*** can support birds and small mammals attracted by the abundant nectar. *Banksia spinulosa* also occurs in woodland on sandstone soils.

▲ An interesting sandy deposit occurs at Agnes Banks south of Richmond, with an almost 'coastal sand-dune' vegetation including the Wallum Banksia *Banksia aemula*, and many other colourful shrubs. Most of this sand deposit has been quarried but a small remnant of the vegetation survives in the **Agnes Banks Nature Reserve.**

Stunted shrubs, spectacular scenery…

Heath — the name conjures up images of desolate windswept expanses stretching to the horizon. Windswept they certainly may be at times, but Sydney heathlands are definitely not desolate, particularly in springtime when they provide some of the most spectacular wildflower displays in the bush.

Heathland plants are shorter than two metres, and shade at least 30% of the ground, but often grow much more densely than this. Most Sydney heath is on sandstone or sand dunes, but it can also occur on clay soils near the sea, where winds and salt spray make growing conditions difficult.

▲ Heath forms where soil conditions are not good enough for trees to grow. Patches of heath on shallow, infertile soil with poor drainage stand out against surrounding woodland in the sandstone landscape.

▲ Strong winds coming straight off the sea keep coastline **heath** plants from growing very tall, as here at North Head.

Variations in local soil, drainage and wind conditions give rise to many different combinations of species grouped under the broad heading of 'heath'. Our Sydney vegetation map describes seven types of Coastal Sandstone Heath, two types of Coastal Dune Heath, as well as Coastal Clay Heath. In the Blue Mountains there is a similar variety.

Coastal Sandstone Heath

▲ A broad rocky ledge is exposed here on a sandstone hillside, and you can see how shallow the soil must be. **Coastal Sandstone Heath** species able to survive here — *Allocasuarina distyla, Banksia ericifolia* and *Hakea teretifolia* — form a dense prickly thicket.

▲ ***Banksia ericifolia*** is a very common and widespread component of heath around Sydney. In autumn its orange flower spikes attract nectar-feeding birds and animals.

▲ Cream flowers appear between the needle-like leaves of ***Hakea teretifolia*** in November. This prickly plant often predominates in locally moister sites.

▲ ***Angophora hispida*** Dwarf Apple grows where conditions are slightly drier, particularly on ridges. The large creamy white flowers attract a diverse array of insects in summer.

Wet heath

Wet heath contains fewer shrubs but abundant grass-like sedges. It occurs on rock platforms where soil is shallow, low in nutrients and poorly drained. These shallow soil patches become very wet after rain, then dry out during drought periods. Plants growing here are able to survive these extreme conditions.

▲ ***Banksia oblongifolia*** is one shrub that can survive intermittently wet conditions and may be found growing with the sedges.

◀ ***Burchardia umbellata***, in the Colchicaceae family related to lilies, develops a corm and tuberous roots in the shallow and poorly drained soil of wet heath.

◀ Common in the wetter heath are species of *Epacris* in the Epacridaceae or southern heath family. In September the upright stem tips of ***Epacris microphylla*** are covered with small flowers and resemble white candles.

▲ Christmas Bells ***Blandfordia nobilis*** are well hidden among sedges until their bright flowers appear. The Christmas Bell family Blandfordiaceae has only four species, three of which are found near Sydney.

▲ In the upper Blue Mountains wet heath is found around the edges of **sedge swamps** formed in valleys with impeded drainage.

▲ One of the most prominent shrubs in sedge swamps and wet heath is the spring-flowering ***Sprengelia incarnata***. It belongs to the southern heath family, Epacridaceae.

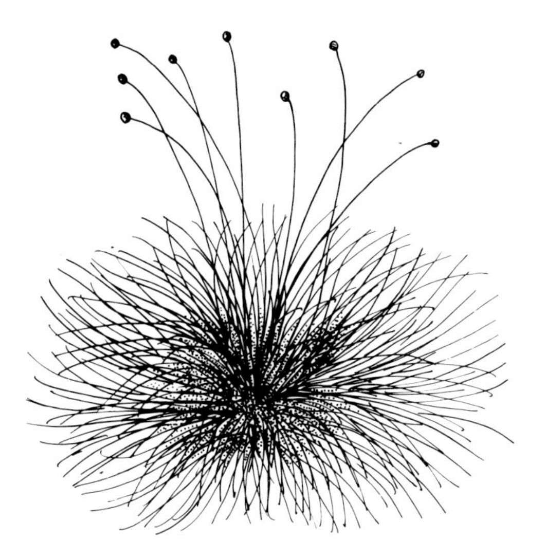

▲ Button Grass ***Gymnoschoenus sphaerocephalus*** is the main tussock-forming sedge in these swamps. It belongs to the Cyperaceae family.

▲ ***Gleichenia dicarpa,*** a Coral Fern, grows along seepage lines. It grows best where there is permanently flowing or seeping water. You may also see it growing on sandstone cliff faces and in roadside cuttings.

Rocky outcrop heath

Where rock platforms are surrounded by shallow soil you will find rocky outcrop heath. The mixture of species varies as it depends on the local habitat, but shrub species in the Myrtaceae family are particularly common.

▲ ***Darwinia fascicularis*** is one of several species of *Darwinia*, in the Myrtaceae family, that can survive in shallow soil on rock platforms.

◄ ***Kunzea ambigua***, also in the Myrtaceae family, may grow over two metres high in shallow soil.

▲ Pink flowers of the smaller ***Kunzea capitata*** are like miniature pom-poms.

▲ ***Isopogon anemonifolius***, in the Proteaceae family, derives its common name, Drumsticks, from the shape of the flowers and seed cones that follow.

▲ Like most ground orchids, ***Glossodia minor*** is inconspicuous amongst the litter for most of the year, with small leaves, or no leaves at all until just before and at flowering time. Leaves and flowers grow anew each year from tubers.

Miniature gardens

Some heath species are able to survive in very shallow soil in depressions in the rocks, forming what look like miniature or bonsai gardens. Mosses often carpet the shallow soil, drying out intermittently but becoming bright green again after rain.

▲ ***Baeckea brevifolia***, with tiny leaves and white flowers, is a common shrub of these rock platform 'gardens'.

Coastal Clay Heath

▲ **Coastal Clay Heath** grows along the coastline north from Long Reef, as soils here have weathered from the Narrabeen Group and contain more clay than soils on Hawkesbury Sandstone. The vegetation is kept low and scrubby by onshore winds, but where there is shelter the heath grades into woodland. *Allocasuarina distyla* is a common shrub.

Coastal Dune Heath

Coastal Dune Heath once covered the sand dunes that stretched from Centennial Park to Botany Bay in Sydney's eastern suburbs. Louisa Meredith enthused as she described its species richness in 1839 — 'Crowded with such exquisite flowers that to me it appeared one continued garden'.

Banksia aemula shrubs are so characteristic of this open-heath that it is called **Eastern Suburbs Banksia Scrub**. The wind-blown sand on which it grows is deep, but leached of nutrients. Organic matter has formed a cemented layer or 'hardpan' a couple of metres or so beneath the surface. This hardpan concentrates water in small soaks in the dune swales, and the swamps that form here contribute to the rich assemblage of species.

Only a few tiny remnants of Eastern Suburbs Banksia Scrub survive in the eastern suburbs, and it has been listed as an Endangered Ecological Community under the *NSW Threatened Species Conservation Act.*

▲ Two distinctive plants of the Eastern Suburbs Banksia Scrub are ***Banksia aemula***, left, and the Grass Tree ***Xanthorrhoea resinifera***, right, with its long flowering spike.

Where the land meets the sea, conditions are hard for land plants

In estuaries, salty water floods the shoreline soil twice a day. Only a few plants have developed ways of surviving these fluctuating, saline conditions. These plants are arranged in bands or zones, related to the height of the tides.

Mangroves are small, twisted trees that grow in the silt that builds up with the ebb and flow of the tides. Mangroves are able to cope with salt and with waterlogged soil with no air pockets. Around Sydney there are two species of mangrove. *Avicennia marina*, the Grey Mangrove, with greyish opposite-paired leaves, is the more common.

▲ Seeds of ***Avicennia marina*** are viviparous, that is, the already sprouting seeds are shed from the tree, to be dispersed widely along foreshores. This hardy, quick-growing plant is able to colonise and stabilise shifting mud and silt that accumulates along tidal shorelines.

▲ ***Avicennia marina*** is widespread on mudflats, but here has established on a sandstone rock platform, with its roots extending into silt accumulated in crevices. The vertical peg-like projections from the mud around the trees are 'pneumatophores', structures that provide oxygen to the roots.

▲ The other mangrove around Sydney is ***Aegiceras corniculatum***, River Mangrove. Its brighter green, alternately-spaced leaves contrast with the greyer leaves of the taller-growing *Avicennia*. In this photo you can see that the leaves are dotted with salt, excreted by the plant as a way of surviving in the saline environment. Clusters of horn-shaped fruits are developing. Where the two species overlap, *Aegiceras* is usually found on the landward or upstream side of *Avicennia*.

Saltmarsh grows on land less frequently flooded by the tides. Plants here are low-growing. *Sarcocornia quinqueflora* and *Suaeda australis* are two of the most common.

▲ Succulent ***Sarcocornia quinqueflora*** may have attractive purplish tints in winter.

▲ Extensive areas of **saltmarsh** remain at Towra Point, on the southern shore of Botany Bay. Elsewhere much saltmarsh has been lost to landfill for sporting fields. *Avicennia marina* trees grow beyond the saltmarsh in the foreground.

▲ On the landward side of mangroves and saltmarsh, where flooding is infrequent, the salt-tolerant Swamp Oak ***Casuarina glauca*** is abundant. The understorey varies with local conditions and may include tussocky *Juncus kraussii* and *Baumea juncea*, as here, or Common Reed *Phragmites australis*. These grass-like plants may also grow where tree cover is absent.

◀ Here ***Phragmites australis,*** a member of the grass family Poaceae, interspersed with the white-flowering salt-tolerant shrub ***Melaleuca ericifolia***, indicate brackish conditions. A band of ***Casuarina glauca*** trees can be seen in the background. This picture was taken in winter, when much of the *Phragmites* had died back because of cold weather.

Salt spray and wind pruning

Beach plants grow in broad zones related to exposure to salinity, but in this habitat the salt arrives combined with wind. Trailing runners of sand colonisers such as *Spinifex sericeus* grass can cope with moving sand and saltspray, and grow on the strandline at the foot of the foredune, immediately above the high tide mark. Back from the sea, coastal dune scrub develops on the foredune.

▲ Salt-laden winds and moving sand make things difficult for plants growing on the beaches.

▲ Coastal Teatree, ***Leptospermum laevigatum,*** grows as a low wind-pruned shrub on dune crests, but can reach the height of a small tree in the shelter of the swales.

◄ Clumps of ***Lomandra longifolia*** grow on headlands and dunes. This hardy plant grows in a wide variety of habitats around Sydney.

◄ Coastal wattles ***Acacia longifolia***, pictured here, and the closely-related *Acacia sophorae*, are usually also abundant on the coastal dunes. *Acacia sophorae* is particularly well adapted to this unstable sandy and windy habitat — its branches trail over the sand like runners.

▲ The trailing vine ***Hibbertia scandens*** grows in sheltered places behind the foredune. Its large yellow flowers are followed by seeds covered in bright orange oily food bodies. Birds and other small animals may disperse the seeds when they eat the food bodies.

▲ ***Banksia integrifolia*** Coast Banksia grows as a small tree in the shelter of the dunes, but is stunted on more exposed dune crests and, headlands. Unlike other Sydney banksias, its cones open every year to release the seeds.

▲ Vegetation on headlands is often subject to severe wind pruning. In very exposed sites with clay soils, as here in Bouddi National Park on the northern side of Broken Bay, trees are absent and vegetation is reduced to grassland with shrubs of *Allocasuarina distyla*, a form of **Coastal Clay Heath.**

◀ Where trees are exposed to wind and salt spray, they are pruned and stunted. Here at Werrong Beach in Royal National Park the dramatic effect of these conditions on coastal rainforest can be seen.

Your feet may get wet!

Wetlands form in swampy places prone to flooding. They fill with water after heavy rain, but dry out gradually afterwards. We find freshwater wetlands in the swales of coastal sand dunes, on river floodplains, and associated with wet heath on rock ledges with shallow clayey soil on sandstone. Soils are quite fertile on the floodplains, but low in nutrients in sand dunes and sandstone landscapes.

Sand dune wetlands

▲ Wetlands form in the swales of sand dunes. In Sydney's eastern suburbs, the extensive system of **sand dune wetlands** formed part of Sydney's early water supply. Now they are all drained or modified. Some, like the one in this photo, persist as lakes in golf courses. *Eleocharis sphacelata* and *Baumea articulata* are common sedge species in these swamps.

Other species that occurred naturally in these wetlands have become popular landscaping plants — Sydney's tallest Paperbark, *Melaleuca quinquenervia,* Crimson Bottlebrush *Callistemon citrinus*, and the yellow-flowered Native Broom, *Viminaria juncea.*

Floodplain wetlands

Wetlands form in the lowest parts of river floodplains, and are most extensive between Penrith and Cattai on the Hawkesbury-Nepean River.

▲ In floodplain wetlands, the lowest-lying parts may be permanently wet, but surrounding bands of soil are alternately wet and dry. Here ***Cyperus exaltatus*** forms a fringe around the margin of Longneck Lagoon in Scheyville National Park, near Windsor. This sedge grows into a large tussock from tiny seeds that germinate when wetland soil dries out between floods.

▲ Another plant that germinates when wetland mud is exposed is ***Nymphoides geminata***, a small trailing plant. Leaves and flowers float when water levels rise. It belongs to the Menyanthaceae family, a worldwide family of aquatic plants.

Many plants that grow in the zone of fluctuating wetness depend on periodically dry conditions to expose mud and stimulate germination of their seeds. These plants are an important part of the biodiversity of freshwater wetlands.

▲ Several species of *Persicaria* grow in this zone in freshwater wetlands around Sydney. ***Persicaria lapathifolia*** (pictured here) is an annual herb that grows up to two metres tall, and can cover mudflat areas very quickly. *Persicaria strigosa* and *Persicaria praetermissa* are lower growing, and die back in winter but resprout with warmth and wet conditions. *Persicaria orientalis* has soft felty leaves and large pink flower sprays. *Persicaria decipiens* can survive drier conditions and is often seen in roadside drainage ditches. *Persicaria hydropiper* grows both in Australia and in southeast Asia, and is known as Water Pepper as it is used to give hot flavours in cooking.

▲ Here are three plants characteristic of swamp margins and streamside habitats. In the foreground, yellow flowers of Water Primrose, ***Ludwigia peploides* subspecies *montevidensis***, can be seen amongst the dark green grass blades of Water Couch, ***Paspalum distichum***. Behind is a tussock of the sedge ***Carex appressa*** — its tough leaves are unattractive to grazing animals.

◄ Swamp Mahogany, ***Eucalyptus robusta***, is another tree often found planted in gardens and parks, but now rare in its natural habitat. It has glossy dark green leaves and large cream flowers that provide nectar for birds and bats in winter. Its natural habitat is the occasionally flooded low-lying land on the edge of floodplain wetlands. Most of this land has been cleared.

BUSHLAND ECOLOGY
— THE WAYS OF WILDFLOWERS

SYDNEY'S BUSHLAND FLORA is like the result of a great natural experiment, intensified by Australia's forty million year isolation as an island-continent. Our bushland contains an impressive assemblage of plants, some from families with ancient lineages whose ancestors were here before the breakup of Gondwana — such as the Wollemi Pine ***Wollemia nobilis*** (opposite) and many of the ferns — others belonging to families that have evolved and diversified in isolation since the breakup of Gondwana, plus occasional long-distance travellers. To this mixture, humans have recently added another imported group — exotic weed species that have become naturalised.

Many of our bushland species, including many endemics — plants that occur nowhere else — have evolved on Sydney's low nutrient soils, and their ecological characteristics today are related to their evolutionary development. These characteristics include structural features like tough leaves, behavioural responses to fire and drought, and specific interactions with animals. Our wildflowers have many diverse ways of living — discovering them is a source of endless fascination.

In plant ecology we study plants in relation to their environment. Ecology is the science that provides the knowledge and understanding necessary for conservation and management of our bushland areas. Sustainability — how the processes of natural diversity keep going — is the key to ecology. Through ecological studies we learn how our natural plant communities function, and how changed conditions and exotic plants may pose major threats to the well-being of our bushland.

◀ Towering trees of the Wollemi Pine, *Wollemi nobilis*, are protected by the steep sandstone walls of this remote canyon. Photo: Jaime Plaza

Ferns and their allies

Plants have evolved through time; we can trace their origins and relationships through the variety of plants we see today. From simple single-celled algae, to the more complex bryophytes — mosses and liverworts — to the 'vascular' plants that develop pipe-like cell structures in which water and nutrients move through the plant. Flowering plants (angiosperms) make up the bulk of the vascular plants in Sydney's bush, but let's start with some of the other plant groups.

The earliest vascular plants reproduced by means of spores rather than seeds, and their descendants include the ferns, and the less highly evolved 'fern allies' — fork ferns, horsetails, clubmosses and quillworts. Wonderful names, aren't they!

▲ ***Psilotum* at the Opera**. Normally found in rock crevices in open-forest or rainforest, the primitive fork fern ***Psilotum nudum*** persists on the sandstone rock face beside the Sydney Opera House.

▲ The 'fishbone' is a common shape for fern fronds. ***Doodia aspera*** is found in various moist forest habitats around Sydney. Frond surfaces feel rough and sandpapery giving it the name Rasp Fern. Here on the right you can see the undersides revealing spores in double rows.

People expect to find ferns in moist places, and ferns do depend on water to complete the life cycle. However, ferns colonise a range of habitats and grow in a variety of shapes — from tiny filmy ferns with fronds a single cell thick, to treeferns as tall as buildings. Spores develop on the underside of fern fronds and their packaging patterns help us recognise the different species.

► In contrast to the widespread Bracken, Rock Felt Fern ***Pyrrosia rupestris*** is restricted to rainforest and moist open-forest. It grows epiphytically on rocks and tree trunks. The small, felty fronds are undivided, and come in two shapes — rounded and elongated — the latter bearing spores on the underside.

▲ Bracken Fern ***Pteridium esculentum*** is one of the ferns with more highly-divided fronds. Bracken is common on well-drained sandy soils, in woodland, forest and coastal heath. The Australian species is closely related to the Bracken of the northern hemisphere — in fact some botanists regard all Bracken as belonging to one species.

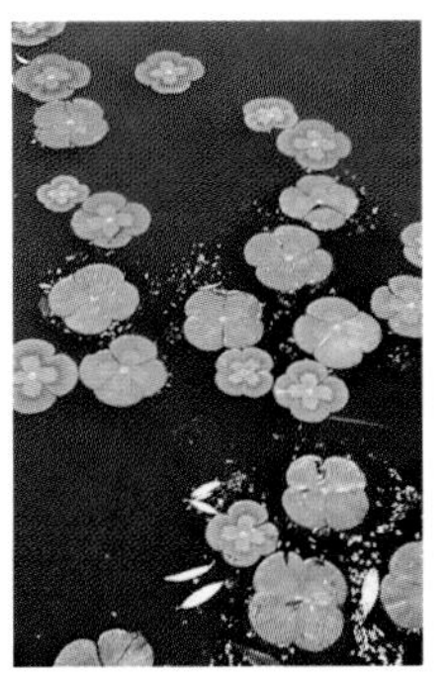

◄ Although ferns grow in moist places, they are not usually thought of as aquatic plants. You could be forgiven for thinking that ***Marsilea mutica*** is a four-leaved clover, but no, it's an aquatic fern, with its roots in wetland mud and leaves floating on the water surface. Aborigines ground the small spore capsules of similar species called Nardoo into a floury edible paste.

▲ Treeferns are at the opposite end of the size scale — you may see them growing metres high in moist eucalypt forests in fertile gullies around Sydney. Their tall trunks are formed from the bases of old fronds. Treeferns like ***Cyathea australis*** feature strongly in forest on basalt caps such as Mt Tomah and Mt Wilson in the Blue Mountains.

Cycads and conifers

Plants with seeds are more highly evolved than the ferns. The more primitive seed plants are the cycads and conifers, most of which have their seeds in cones; none have flowers.

▲ ***Macrozamia communis*** with its palm-like leaves is not a palm but a **cycad**. Known as Burrawang, it is often a conspicuous plant in coastal forests on sand dunes. The large seeds provided food for Aborigines — after treatment for several days to leach out poisons.

Conifers are trees or shrubs. Their seeds generally form in cones. Pollen blows from male to female cones, but as they don't have flowers, there are no flower parts to protect developing seeds (hence the old botanical term, gymnosperm, meaning naked seed, which applied to cycads and conifers).

Our most abundant conifer genus is *Callitris*, in the family Cupressaceae — the native Cypress Pines. These trees are common in rocky or sandy parts of inland New South Wales, but are only a minor component of bushland around Sydney. *Callitris muelleri* and *Callitris rhomboidea*, the Port Jackson Pine, are widespread but not common — *Callitris rhomboidea* may be seen, for example, on rugged sandstone hillsides along the lower Hawkesbury River. *Callitris endlicheri*, Black Cypress Pine, is found in the upper Blue Mountains but is more common further west.

▲ The metre-high ***Microstrobus fitzgeraldii***, in the conifer family Podocarpaceae, is found only in the Upper Blue Mountains in the spray zone of a limited number of waterfalls between Wentworth Falls and Katoomba. Its restricted occurrence may have arisen for similar reasons to that of the Wollemi Pine.

The Sydney region's most spectacular conifer by far is the Wollemi Pine *Wollemia nobilis*, in the family Araucariaceae. This 40-metre tree aroused worldwide excitement when discovered north-west of Sydney in 1994. It is similar to plants that are now known only as fossils from the Jurassic Period, the 'age of the dinosaurs'. Yet a tiny population of *Wollemia* has survived in a remote canyon in the rugged Wollemi National Park. Perhaps this moist, sheltered area has provided protection from fire or from periodic drought as the climate fluctuated over millions of years.

Conifers related to *Wollemia* were abundant in ancient forests until about 40 million years ago, but have slowly declined and been replaced by flowering plants. *Wollemia's* present-day relatives, tall *Agathis* and *Araucaria* trees like Kauri Pine and Norfolk Island Pine, are found in the rainforests of northern New South Wales and Queensland, on islands of the western Pacific Ocean and in Indonesia, and in South America.

Flowering plants developed later in the evolutionary sequence and are known as **angiosperms** (meaning enclosed seeds). Flowering plants outnumber the other vascular plants both in the bush and in this book — and of course they are generally more noticeable because of their flowers, colourful structures which attract the attention of pollinators as well as bushwalkers.

▼ Male and female cones of ***Wollemia nobilis*** are borne on the same tree, but on the tips of separate branches. Female cones are round, and the male cone, shown here, is cylindrical.

Eighty million years ago the world of the dinosaurs was splitting apart …

Eighty million years ago the supercontinent Gondwana that had connected Australia through Antarctica to South America and Africa, was splitting apart. Pressures from within the earth's crust were breaking up the world of the dinosaurs. Africa and South America separated earlier from Antarctica. Australia, still joined to Antarctica, moved slowly northwards, finally separating from Antarctica about 40 million years ago.

The occurrence of related plants in widely separated continents had puzzled botanists for many years. The flowering plant families Proteaceae, Myrtaceae, Casuarinaceae and Restionaceae are common around Sydney, and like the conifer family Araucariaceae, have many relatives in other Southern Hemisphere continents, particularly southern Africa and South America. As recently as forty years ago these occurrences of related species in distant continents were explained by long distance dispersal of plant seeds between continents.

▲ Migrating birds may distribute seeds and propagules over long distances. Many wetland plant species are widespread probably due to movement by waterbirds. The soft-leaved herb ***Persicaria decipiens*** occurs naturally in wetlands along the Hawkesbury as well as in Eurasia and Africa. Its small seeds may be carried in mud attached to migrating waterbirds.

◄ Fern spores are small and readily dispersed by wind and upper air currents and this may explain the widespread or cosmopolitan distribution of Australian species such as the Common Maidenhair Fern ***Adiantum aethiopicum***. Its soft, delicate fronds make it popular as a potted plant, but it grows naturally in Sydney forests.

The discovery that the continents actually drifted or moved, taking their plants for a ride, provided a breakthrough in explaining the distribution of the southern (Gondawanan) families such as the Araucariaceae, Proteaceae, Myrtaceae, Epacridaceae, Casuarinaceae and Restionaceae.

Alone, and adrift

Since its separation from Antarctica about 40 million years ago, Australia's flora and fauna have been evolving in isolation from the rest of the world. As the island continent drifted northward the climate became unstable and at times very arid. During climatic fluctuations including 'ice ages' over the past 25 million years, the ancestral closed-forests of Gondwana have been reduced periodically to 'refugia', pockets of forest persisting where local conditions of rainfall and geography allowed. The position of these refugia does not appear to have changed, though the forest areas have contracted and expanded in response to the climatic fluctuations. The last 'ice age' or glacial maximum was just 18 000 years ago. Sydney's climate has been relatively stable for the last 6000 years.

▲ Distribution patterns for some groups of plants are difficult to explain without the concept of continental drift. The Waratah ***Telopea*** has closely related cousins also belonging to the Proteaceae family — species of *Protea* in South Africa and other related species found only in South America. Waratah seeds are only locally dispersed. Long distance dispersal seems inadequate to explain these distribution patterns.

► Representatives of plant families such as the Fabaceae, Asteraceae and Poaceae are found throughout the world. However, many species are endemic — naturally restricted — to Australia. For example, most of the 700 or so Australian species of wattles belonging to the genus ***Acacia*** are endemic. The high proportion of endemic species is the result of Australia's long period of isolation after the break up of Gondwana. About 90 species, including ***Acacia ulicifolia***, are found around Sydney.

The predominantly hard-leaved sclerophyll families, Proteaceae, Myrtaceae, Epacridaceae, Casuarinaceae and Fabaceae have evolved on Australia's low nutrient sandy soils. In contrast 'rainforest' families such as the Sapindaceae, Meliaceae and Lauraceae, containing mostly soft-leaved, moisture-requiring plants, occupy the more restricted higher nutrient soils.

Present-day distribution of vegetation, patterns of rainforest, forests, woodlands and heaths, and their particular composition, are the result of historical processes, limited by the constraints imposed by the current climate. Metaphorically speaking, today's plants have been sifted through the twin sieves of natural selection and climatic change from a flora developed over millions of years. The human impact of clearing has, of course, drastically reduced native vegetation, particularly on fertile soils.

Naturally restricted and isolated populations of some species appear to be the result of the sifting process. For example, the Wollemi Pine *Wollemia nobilis* was formerly more widespread but has become much more restricted as climate has changed over thousands and millions of years. Of course such species may expand again if conditions — such as increased rainfall — become more favourable. Species are not destined to become extinct just because they are rare. It is current human activities, such as clearing, over-collecting seed or the accidental introduction of pathogens, that are more likely to cause extinction.

Coping with changing conditions

Why are plants so many different shapes, sizes and colours? In nature, this variety is the result of evolution by natural selection over long periods of time. It involves changes to individual plants that happen by chance from one generation to the next, affecting different processes in the plant. These changes end up helping or hindering the new plant's survival in its environment. By natural selection the more successful variations are passed on to subsequent generations.

The plants we see around us today are the survivors. Of course the plants haven't consciously planned these changes, but it's hard to avoid speaking about them in human terms. These helpful changes are like survival tactics — we call them **adaptations** to the environment.

We can see adaptations in all plant parts — leaves, stems, flowers, roots, tree trunks. There are also adaptations we can't see — for example, in a plant's internal chemistry. Adaptations in response to one set of conditions may later help the plant survive a different set of conditions. For example, resprouting that helps plants survive fire may have originally helped plants survive drought.

Leaf shapes and substitutes

Species of *Acacia* illustrate variations in the basic plant form that have been successful.

◀ Like most wattle species, Sydney Golden Wattle ***Acacia longifolia*** has, instead of true leaves, leaf stalks flattened to look and act like leaves. These 'phyllodes' come in a variety of shapes and colours, but all contain green pigment. They can photosynthesise like leaves, but lose less water while doing it. This an adaptation to dry conditions.

▲ A few acacias, like the Black Wattle ***Acacia mearnsii***, have true leaves, but these are divided into small segments. These 'bipinnate' leaves lose less water than would a similar-sized undivided leaf.

All *Acacia* species begin life with true leaves and those with phyllodes develop these as they grow older. You may see seedlings with both, in the process of changing.

Not in good taste

Eucalypts of the Myrtaceae family are another major and characteristic part of the Australian flora. There are almost 1000 species; about 80 species are found in the Sydney region, in almost all habitats. Eucalypts have kept true leaves, but their leaves are modified in ways that help save water and protect them from being eaten.

▲ We can see examples of leaf modifications in ***Eucalyptus bicostata***, found naturally near Jenolan Caves. Its young leaves are different from the adult leaves. Juvenile leaves have a white waxy coating that helps protect the young tree against frost. Adult leaves hang vertically to reduce exposure to the sun, so minimising water loss. Thick waxy coatings that give the leaves a shine also protect them from drying out.

You can't see it in the photo, but all *Eucalyptus* leaves contain oil. Eucalyptus oil may smell nice to us, but it tastes bad enough to deter many would-be munching insects and larger animals. Aromatic oils occur particularly in the Myrtaceae — for example in species of *Eucalyptus, Leptospermum* and *Melaleuca* — and in the Rutaceae or citrus family, including species of *Boronia, Eriostemon* and *Zieria.* You can see oil-producing glands by holding leaves up to the light.

▲ Oil glands in the leaves are visible as white dots in ***Boronia floribunda,*** like the dots in leaves of orange and lemon trees which also belong in the Rutaceae family. *Boronia floribunda* is a beautiful but uncommon shrub of heath and open-forest in Sydney sandstone areas.

Nasty-tasting sap is another way of protecting leaves from being eaten — this is common in species of fig, *Ficus* in the Moraceae family, and climbers of the Apocynaceae and Asclepiadaceae. The garden Oleander in the Apocynaceae has a well-known poisonous reputation.

▲ Giant Stinging Trees ***Dendrocnide excelsa***, found on rainforest edges, could be considered to have the ultimate in tactical defence. Like nettles, which belong to the same family, Urticaceae, their stinging hairs, especially numerous on larger juvenile leaves, repel even botanical collectors.

Sclerophylly — a typically Australian response

The sandy, low nutrient Sydney sandstone soils make growing conditions tough for plants. Because the supply of nutrients to make new leaves is limited, plants can't 'afford' to lose leaves in drought, or have them eaten. Many have adaptations that protect against these hazards.

Leaves of many species contain fibres (sclereids) that act like the reinforcement that strengthens concrete — they keep the leaves from wilting when it's dry, they also make them unpalatable to eat. We call these leaves ***sclerophyllous***, meaning hard-leaved. Sclerophyllous species are common in the families Proteaceae, Myrtaceae, Epacridaceae, Casuarinaceae and Fabaceae.

▲ ***Styphelia tubiflora*** is one of many Sydney sandstone species in the Epacridaceae or southern heath family that have small tough sclerophyllous leaves.

▲ ***Hakea sericea*** in the Proteaceae family is a prickly shrub of the sandstone woodland. Its leaves are like needles — their shape minimises water loss, and the sharp point deters munching animals. The thick rough fruits may have developed woodiness originally to protect the plant's seeds from being eaten, but now protect the seeds from fire. Cockatoos, with their strong beaks, can break into the fruits.

In the Casuarinaceae or Sheoak family, evolution has reduced leaves to tiny tooth-like scales, visible in circlets around branchlets. The branchlets, called 'cladodes', are green and do the photosynthesising for the plant, in a similar way to the phyllodes of *Acacia.*

▲ Sheoak flowers are minimal too. Pollen-shedding anthers are the most conspicuous parts of the male flowers, seen here arranged in circlets in the very rare ***Allocasuarina portuensis.*** Male trees when flowering have a burnished brown appearance. Like grasses, sheoaks rely on pollination by wind.

▲ Here you can see the female flowers of ***Allocasuarina distyla*** — they appear as red bunches, which are mostly the stigmas or parts that catch pollen. Below is the woody fruit cone that develops, protecting the winged seeds from fire and most birds — except Glossy Black Cockatoos that chew through the wood.

▲ Tough sclerophyllous leaves are so common in the sandstone landscape people often assume a soft-leaved plant is an introduced weed. ***Astrotricha floccosa***, a straggly shrub which grows towards the base of sandstone hillslopes is often mistaken in this way. In this sheltered environment it can survive with soft leaves. However, its large soft leaves are much more attractive to munching animals than small hard sclerophyllous leaves, and they are protected with a thick coating of hairs on the underside. In fact the name *Astrotricha* means 'star-shaped hairs', and *floccosa* means 'woolly'.

▶ Some plants have dispensed with leaf-like structures completely. ***Cassytha glabella*** (like its more robust relative *Cassytha pubescens*), in the Lauraceae family, has tiny leaves like scales on thread-like stems — it is semi-parasitic, attaches to stems and leaves of other plants by small pads called 'haustoria', and absorbs their nutrients. Its fleshy fruits are spread by birds, and seedlings germinate in soil. Once the growing stem has attached itself to another plant, the connection with its own roots dies, and it spends the rest of its life trailing over other plants, sometimes forming dense mats.

A land scorched by fire

▲ 'Bushfire destroyed thousands of hectares of bushland …' trumpeted the media headlines. But a month later, on the blackened ground, amidst grey ash, dead sticks and charred tree trunks, green shoots are appearing. A year or two later, the whole scene will be green. Bushfires are a dramatic but natural part of the Sydney landscape. Fire intensities vary, and even after large extensive fires there are small patches that have been missed. Here despite the intense scorch, ***Xanthorrhoea resinifera*** Grass Trees are quickly resprouting. Amongst the inner leaf bases you may find small insects and spiders that survived the fire by taking refuge there.

► Most eucalypts can resprout from trunks and branches. This old Scribbly Gum ***Eucalyptus haemastoma*** is a veteran of many fires. A month after the most recent one, new shoots are already emerging from its trunk.

◀ Even though their stems are killed by fire, many woody shrubs resprout from roots or stem bases. Dwarf Apple ***Angophora hispida*** resprouts from lignotubers — woody swellings, partly or wholly underground, that contain buds.

Rising from the ashes

Not all species can resprout after fire. Plants of some species are killed and a new generation must begin from seed.

▲ Along with the flames, fire produces heat, smoke and ash. Each of these may be advantageous to plants re-establishing after fire. Heat may open woody cones or capsules, smoke may stimulate seeds to germinate, and ash may act as a fertiliser.

◀ ***Petrophile pulchella*** seeds may survive in woody cones or capsules that open after the heat of the fire. Emerging seeds have hairs that catch the wind and help anchor the seeds when they land a short distance away on the bare ground.

▲ Thick woody cones like those of ***Banksia oblongifolia*** provide good protection from fire and the heat causes them to open afterwards. Winged seeds are shed onto the nearby bare, ash-enriched soil surface, ready to germinate after rain. On sloping ground, rain rearranges fallen scorched leaves and twigs into 'litter-dams' that may catch seeds and protect them from being washed away.

Some species, such as *Banksia ericifolia,* are killed by fire, and seeds released from its cones are needed to provide the next generation of seedlings. **But what happens if there's another fire before the seedlings have grown to maturity?**

◀ Only a small proportion of the seeds a plant produces will result in adult plants. This seven-year-old seedling of ***Banksia ericifolia*** has just produced its first flower. It may be ten years old or more before it has produced enough seeds to guarantee a new generation. A fire before then may kill off the plants without leaving any seed to survive.

Most Sydney plant species can survive bushfires provided fires don't occur too often. Our bushland plants and animals have evolved with intermittent fires over millions of years.

Bunkered beneath the soil

What about plants that don't store seeds in woody cones, but shed them onto the soil each year? How do these seeds avoid being burnt in the next fire, or being eaten in the meantime?

Many of these species have seeds with hard coats and attached 'food bodies' — these are attractive to some types of ants. The seeds become buried in the ants' nest after they eat off the food body. The seeds become distributed at different depths within the soil, so when a fire passes above they are protected from being burned and their dormancy is broken by the heat or smoke. The seeds later germinate after rain, and seedlings grow in full sun in the bare post-fire soil, enriched by ash.

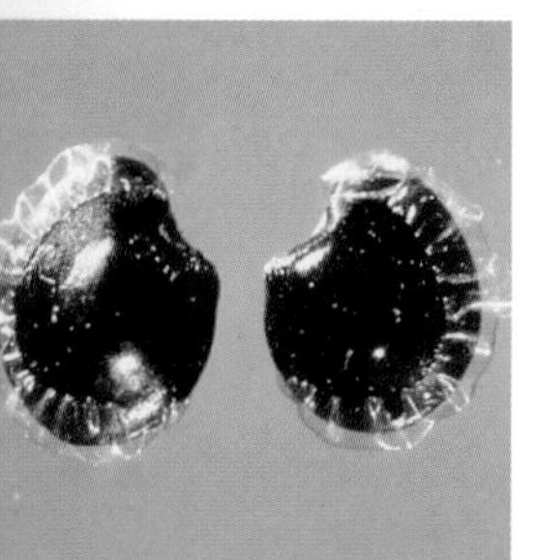

◀ Many shrubs and herbs, particularly in vegetation on infertile sandy soils, have seeds with hard coats and ant-attracting food bodies. The food bodies on ***Dodonaea triquetra*** seeds look like tiny shower caps. Soil-stored seeds germinate after fire.

▲ Regeneration from seed depends on post-fire weather conditions, which may vary widely. For example, seed of the beautiful rare Native Rose ***Boronia serrulata*** also has a food body and becomes stored in the soil. One population didn't regenerate after a fire until heavy rain stimulated soil-stored seed to germinate 18 months later.

Smoke stimulates many seeds to germinate. For example, species of *Dampiera, Scaevola, Eriostemon, Hibbertia, Conospermum* and *Grevillea* have been found to respond in this way. Research on responses of Sydney species is continuing, after smoke was found to help germination of wildflower species in South Africa and Western Australia.

▲ Grey Spider Flower ***Grevillea buxifolia*** occurs only in the greater Sydney region where it is common in woodland on sandstone. Smoke improves germination of its seeds by 50%.

Fire is an important part of the bushland cycle. Without fire the sclerophyllous species that have adaptations to survive and regenerate after fire may be outgrown by species such as *Pittosporum undulatum* and by introduced weeds that establish in the absence of fire. Many small urban bushland reserves have dense *Pittosporum* canopies because of the absence of fire.

▲ Despite their proximity to the city it is important that bushland areas are occasionally burnt. Indeed fire seems necessary to maintain local populations of many plant species. Though the heath here near Dobroyd Head on Sydney Harbour appeared to have been completely destroyed by a hot fire in November 1990, it has recovered well, and is enjoyed by people using the Manly Scenic Walkway.

However it's important for conservation of all our bushland species that fires are not too frequent. People often ask — what is the 'correct' frequency for fires? The answer is — it varies for different types of vegetation, and it varies for different species. When our bushland was intact, natural and Aboriginal-induced fires resulted in a patchy mosaic of long-burnt, recently-burnt, unburnt and long-unburnt bushland, at different growth stages after fire. If a species failed to survive two fires in quick succession in one place, it was likely to be still present close by, and be able to recolonise from there. Nowadays this no longer applies as our bushland is much reduced in area and often highly fragmented.

◄ Rare plants that occupy a very localised habitat and that are killed by fire, like the beautiful ***Grevillea caleyi***, may be particularly vulnerable to frequent fires.

Invisible ecology — what's happening underground?

Hidden from sight in the soil, the roots of bushland plants carry on secret 'liaisons' with soil organisms. A well known example happens in the pea family Fabaceae. *Acacia* and other peas develop nodules on their roots. Bacteria live in these nodules and extract nitrogen gas directly from air pockets in the soil and incorporate it into compounds the plant can use. These beneficial bacteria are the 'nitrogen-fixers'.

▲ In the roots of shrubby ***Acacia elongata*** are nitrogen-fixing bacteria that help to supply them with nitrogen. This is in short supply in the sandy soils where these shrubs grow.

Less well known is that most of our bush plants have associations with fungi. Mushrooms and toadstools are the above-ground, visible fruiting bodies of many fungi, but the major portion of fungal growth occurs below ground as tiny hairlike 'hyphae'. These grow through soil, rotting wood and litter, breaking down dead material and collecting the nutrients.

Mostly we think of fungi as causing diseases in plants rather than being beneficial. But there's a group of 'friendly fungi' that team up with plant roots to form **mycorrhizas** — literally an association of a fungus, 'myco', with a root, 'rhiza'. These fungal hyphae grow around or into a plant root and tap into the plant's food supply that has been manufactured by photosynthesis high up in the leaves. In return, the fungus transfers to the plant some of the raw materials needed for food production — nutrients absorbed from decomposing matter in the soil by the hyphae. The hairlike hyphae have a large ratio of surface area to volume, and so are very efficient at absorbing soil nutrients. For one centimetre of root associated with mycorrhizal fungi there can be up to three metres of hyphae! The plant partner receives mineral nutrients from a much larger volume of soil than its own roots have penetrated. The association is beneficial for both parties — particularly in areas of low-nutrient soils, such as the low-phosphorus soils on Sydney's sandstone.

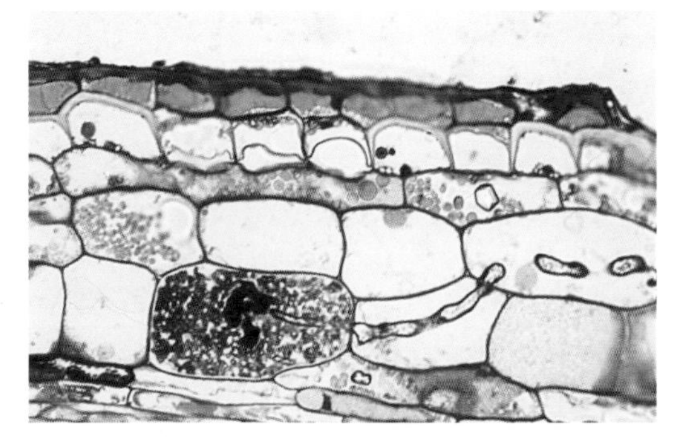

▲ You can see the **mycorrhizal association** between fungal and plant cells in this cross section of a Wollemi Pine root photographed through a microscope. The root cells are stained green and fungal tissue is stained purple. One root cell towards the bottom contains a lot of fungal tissue, which provides a large surface area for the transfer of nutrients and food between plant and fungus.

Mycorrhizas may also help plants, especially seedlings, growing in deep shade under dense canopies, as Wollemi Pine seedlings do.

▲ In deep shade in sandstone gullies, seedlings of Coachwoods, ***Ceratopetalum apetalum***, have been found to receive food manufactured in the canopies of nearby large adult trees — via interconnected mycorrhizal hyphae. Apparently these seedlings have tapped into the fungal internet!

► Another underground tactic to obtain nutrients is stealing from other plants. The roots of Native Cherry ***Exocarpos cupressiformis***, in the Santalaceae family, tap into roots of other plants to supplement their food and water supplies. There does not seem to be any deleterious effect on the host plants, which may include eucalypts.

Plants without partners

◄ Plants in most families can form mycorrhizal associations. However, plants in the family Proteaceae, which are so prominent on low nutrient soils in Sydney bushland, instead form **proteoid roots** — roots with outgrowths that increase the surface area for nutrient absorption.

▲ Proteoid roots form mats as they grow linked together and up into leaf litter. As well as helping plants gain soil nutrients, proteoid root mats may help plants by protecting soil from erosion after bushfires. In recently burned bushland you may see proteoid root mats protruding from the bared soil surface, near *Banksia* shrubs for example.

At home among the gum trees

Long-suffering plants

Most encounters with animals end with the plant being eaten — kangaroos eat grasses and small herbaceous species, koalas eat *Eucalyptus* leaves, possums and flying foxes eat fruits, leaves and nectar, many birds eat seeds and fruits.

◀ Insects are small but important consumers of plants. ***Eucalyptus amplifolia*** on low-lying floodprone land is called Cabbage Gum because of its large juvenile leaves. Like garden cabbages, these leaves attract a variety of insects. Native insects make major contributions to bushland's biodiversity.

▲ The small lumps on these eucalypt leaves are the homes of tiny insect larvae, which modify the leaf tissue into distinctive '**galls**'. The adults emerge and lay eggs on new leaves. There is a wide range of galls made by different insect species, particularly wasps and flies, on different plants and plant parts. Some large galls may be almost as big as golf balls. Gall-forming insects rarely cause severe damage and are part of our bushland's ecology.

◀ Plants also provide important wildlife habitat. Trees and shrubs provide small birds with cover from predators, and also provide perches and nesting sites for larger birds, such as **Sulphur-crested Cockatoos** which nest in tree hollows. Trees may take many years to develop holes suitable for nest hollows and the loss of old habitat trees in urban and rural areas leads to a gradual loss of animal biodiversity.

Win-win encounters — *'of birds and bees'*

There may be more equitable relationships between plants and animals. Many plants provide food to entice an animal to do something for them in return.

pollen to the next flower. The introduced Honey Bee is well known in this role, but it may harm native flowers by taking nectar without pollinating them and so making the flowers unattractive to other potential pollinators.

▲ Red may alert humans to danger, but to birds it alerts them to food. The nectar from the red bottlebrush flowers of ***Callistemon linearis*** provides food for honeyeaters. As they take the nectar they inadvertently pick up pollen which they carry to the next flower, so pollinating it. Possums, gliders and flying foxes use plants as food sources and are important plant pollinators.

Other plants attract insects that provide a pollinating service. Plants attract insects to their flowers by means of colour and scent, and provide food for the insects in the form of nectar. In return the insects pick up and carry

◄ Some plants have developed elaborate mechanisms to ensure they attract an insect that actually achieves pollination. The stigma — or pollen-collecting pad — of the Triggerplant ***Stylidium productum*** flower is on a moving 'arm', like a booby trap that springs when an insect probes into the flower to drink nectar. Pollen on the insect's back is thereby transferred to the stigma.

Ants and confidence tricks!

Having achieved pollination and produced the next generation of seeds, how can parent plants get rid of their offspring? Some plants encourage ants to help them. Their seeds have an inviting, oil-rich food morsel attached to them. Ants of particular species carry the seed back to their nest, where they eat the food body and discard the rest of the seed, either underground in a back corner of the nest, or on the rubbish heap outside. A hard seed coat generally helps protect the seed embryo from being eaten, and from rotting in the soil.

◄ Here a seed of ***Acacia terminalis,*** with its sickle-shaped food body, is being manipulated by ants towards their nest.

Plants with ant-attracting food bodies on their seeds occur widely in Sydney, but are most common in woodland and heath on sandstone. In these habitats up to half of the shrubs and smaller plants have seeds with food bodies. Plants include peas and wattles, in the Fabaceae family, and species of *Boronia, Eriostemon, Zieria, Grevillea, Hibbertia, Dodonaea, Bauera, Leucopogon, Dampiera, Scaevola, Tetratheca, Actinotus, Platysace, Xanthosia, Ricinocarpos, Patersonia, Caesia, Thysanotus, Restio, Lepidosperma* and *Lomandra*. Heat and smoke from fires stimulate germination of the soil-stored seed.

◄ Ant-attracting food bodies attached to seeds take a variety of shapes. Here is a small selection from woodland plants. Other food bodies include nets, flaky coatings, 'shower caps' and oil-filled 'quilts'.

Sex and the single wasp

Not all plant-animal interactions benefit the animal. Plants can sometimes trick animals into pollinating them without giving them anything in return, except embarrassment!

▲ This Purplish Beard Orchid ***Calochilus robertsonii*** is not trying to resemble a 1960s hippy — it's trying to resemble an insect! The shape of its flower attracts an unsuspecting wasp keen on mating with another wasp. By the time the, presumably embarrassed, wasp realises its mistake, the orchid has been pollinated.

Turning the tables!

Some plants have turned the tables on animals completely. Sundews — small herbs rarely taller than 50 cm — are carnivorous plants that 'eat' small invertebrate animals.

▲ ***Drosera binata*** is a carnivorous sundew — it traps small insects with sticky hairs on its leaves and then digests them. In this way, sundews gain nutrients, especially nitrogen, that are hard to extract from the waterlogged soils where they grow. Its attractive regrowing leaves make this sundew particularly noticeable after fire.

▲ Another carnivorous sundew is the smaller ***Drosera peltata***. The insect-catching hairs are prominent on leaves which alternate up the stem. Luckily, human-eating sundews are scarce in Sydney bushland!

◀ *Utricularia* or bladderworts — some are called Fairy Aprons — are another group of plants that live in wet places and supplement their diets with animals. They catch tiny soil-dwelling nematode worms in little traps. Here on the Tasmanian Bladderwort ***Utricularia monanthos*** the traps are on root-like stems.

Live and let live

Enjoy the bushland for all its wildlife. Humans are the greatest hazard for most wildlife — either directly through destruction of habitat, or indirectly through pollution of catchments or by allowing pet cats and dogs to roam in bushland.

▲ Lizards such as this **Water Dragon** are frequently met with in the bush. Enjoy the experience but please don't try to catch or harm them.

You should have seen the flowers here last spring!

Seasonal change in mainland Australia is not accompanied by the major landscape changes that snowfalls and deciduous leaf fall bring to other places. Some light winter snowfalls may be experienced in the Blue Mountains if you are lucky, but these seldom last more than a few hours. In Sydney's midwinter **July** it's cold, but wattle flowers begin to burst open — by **August,** blazes of golden colour are scattered across the landscape.

▲ The end of winter brings on the most spectacular of the wildflower displays, and by **September**, the spring wildflower display on the sandstone is in full swing. The yellow of *Acacia* and other peas vies with the pink of *Boronia* and *Eriostemon* as the dominant colour.

▲ In **September**, in woodland on shale soils, flowering stems of dainty lilies and orchids appear seemingly from nowhere, in reality sprouting from bulbs and tubers. Daisies, native geraniums, and native bluebells — species of ***Wahlenbergia***, above — all appear in the understorey.

Summer, **December – February,** is the time of greatest seed production, but there are still many plants flowering. The tree banksias, *Banksia serrata* and *Banksia integrifolia*, bear spikes of flowers, and small blue flowers of *Dampiera stricta* appear close to the ground. In shale woodland, lilies, orchids and small herbs shrivel and die back to rootstocks in the summer heat, and it is the turn of grasses to flourish.

▲ Dry summers may lead to bushfires, such as here in the shale woodlands of western Sydney. A single fire doesn't 'destroy' the bush, because plants have evolved along with hot summer fires over many years.

In autumn, **March – May,** the results of fungal activity become visible — this is the time you are most likely to see mushrooms and toadstools — the visible fruiting bodies of fungi.

◄ Autumn is also the time to see spectacular shrubby banksias like ***Banksia marginata***, their solid flowering spikes no doubt a welcome sight to honeyeaters flying through the sandstone woodlands. The summer grasses die back and there are lots of small white flowers in woodland on sandstone. On still days the woodland air may be heavy with sweet scent from prickly *Woollsia pungens* bushes.

◀ **June – August**, winter again, and you may be delighted to find ground orchid flowers that have popped up amongst the leaf litter. Here is the Trim Greenhood ***Pterostylis concinna***. Wattles are now heavy with buds ready to burst open again in July and August.

Should we blame El Niño?

Superimposed upon the annual seasonal changes are major events — floods in wetlands and fires in bushland — that are likely to happen episodically, only every few years. The ecology of our bushland plants is driven by changes brought about by these episodic events. Floods and fires kill many plants — but they also provide conditions for germination and renewal. For plants, these events are not 'catastrophes'. Plant survival strategies — resprouting and seeding — are similar for both.

Regional weather cycles over decades are as important for bushland plants as the seasonal cycle. The annual cycle is important in regulating flowering and seed set, but seedling establishment and growth may be more in tune with the alternation from the wet — brought by La Niña — to the dry of El Niño. Periods of growth over several years are followed by drought and fire. Rather than the regularity of the seasons, it is the variability over decades that seems to make sense as the driving ecological force in the Australian context.

Seasonal change versus episodic events

In freshwater wetlands you're likely to see some of the most dramatic seasonal changes, as well as contrasts from year to year. The following pictures of Reedy Swamp in Cattai National Park show the same scene at different times.

▲ In March 1992 water levels in Reedy Swamp were high after big floods, but a clump of the sedge *Schoenoplectus validus* was visible.

▲ By October the lush growth that followed the floods had died right back during the cold dry winter, exposing cracked mud. This bared surface provided ideal conditions for germination of many wetland seeds in the mud.

▲ Eighteen months later, in autumn the formerly bare mud is covered with lush growth of ***Cyperus exaltatus,*** a large tussocky sedge that grows from tiny seeds. These plants die back in winter as part of the seasonal cycle.

A changing flora?

Our bushland plants have evolved in relative isolation from other continents over millions of years but humans can move plants from one end of the globe to another in less than a day. **How does the introduction of plants that have evolved under different conditions elsewhere affect our bushland?**

There is a group of plants that appear and spread quickly in new places. If you have ever had a garden, you will know these plants — they are the weeds! In your garden or farm, you dig the soil and add fertiliser — and the weeds appear!

Of course, the weeds don't appear from nowhere. Initially they've been introduced to Australia by humans — either accidentally or deliberately as food, fibre or ornamental plants. Many of these deliberate introductions have remained in the garden or pasture but some have escaped and spread into bushland.

There are almost always weeds invading bushland from the edges of farms, gardens, parks and roadsides — **why?** At the edge of these places, disturbance clears space and makes light available for weeds to germinate and grow, then watering brings extra moisture and fertiliser in runoff to the next-door bush. All this makes a situation similar to a prepared garden bed — with cleared soil, extra water and soil nutrients — a habitat that weeds are well adapted to colonise! Their seeds arrive in the wind, in runoff water, in dumped garden clippings, or in droppings of birds or animals. Once established, weeds are programmed to grow faster than most native plants.

Much bushland on the edges of farms and suburbs, and along the edges of creeks, has a large complement of weeds. They come in a variety of shapes and sizes, from small creepers to large shrubs and trees. But they are mostly at the edges. That's why some people think the bush is untidy — because they see the edges that the weeds have messed up.

▲ Large-leaved Privet ***Ligustrum lucidum*** was introduced as a hedge but is now one of Sydney bushland's most troublesome weeds. Like most weeds, it allocates a lot of energy to reproduction! These fleshy fruits are spread by water and birds. In nutrient-enriched soil, seedlings can grow faster than most native plants.

More nutrients? Please don't feed our bush!

Our sandstone soils are very infertile — and not generally a good place for weeds to flourish. But bushland here is vulnerable to weed invasion in two places — at edges of cleared land, and in gullies along creeks. In our sandstone landscape most development is on the ridgetops, but its products — extra moisture, fertiliser in runoff water, and weed seeds — flow downhill and make an environment favourable for weed growth in moist gullies.

▲ This creek at the edge of suburban bushland has all the ingredients for weediness — development of the catchment, disturbance at the edges, nutrient-enriched runoff. At least a dozen weed species are visible here. Improved catchment management and bush regeneration work are needed to restore it.

▼ Compare the mess in the previous scene with an undisturbed, weed-free creek in Royal National Park. Wouldn't we rather have bushland creeks that look like this?

▲ **Friend or foe?** Sweet Pittosporum ***Pittosporum undulatum*** is a small tree native to the understoreys of Blue Gum High Forest and Turpentine-Ironbark Forest, most of which have been cleared. It has been called a weed by some people. But in Sydney it's an indigenous plant that has benefited from the changed conditions at bushland edges — increased moisture and fertiliser in runoff, plus the absence of fire. Mature trees tend to suppress growth beneath them, and may suppress weeds at bushland edges.

Native and weed look-alikes

So you want to help 'weed' the bush!

Before you start, it's important to know the natives from the weeds. You can learn more about bush regeneration, weeds and how to keep our bushland in good condition, at a TAFE College, for example. One thing students learn is how to distinguish between native and weed look-alikes. Here are a few examples.

Look-alike small trees

▲ The NATIVE Grey Myrtle ***Backhousia myrtifolia*** is a bushy small tree with smooth grey bark that grows along creeks and rivers and may be confused with *Ligustrum sinense*. Its opposite leaves have distinctive veins and oil dots that are visible when held up to the light. We've seen it felled in mistake for privet. It may be confused with:

◄ The WEED Small-leaved Privet ***Ligustrum sinense*** shares white flowers, opposite leaves and gully habitats with *Backhousia myrtifolia*, but its flower heads are cone-shaped, and leaf margins often undulating. Purplish-black fruits hang in dense bunches in winter. Care is also needed to avoid confusing this weed with the native shrub *Breynia oblongifolia*.

Look-alike small trees

▲ The NATIVE Cheese Tree ***Glochidion ferdinandi*** has alternate leaves, paler underneath like those of *Ligustrum lucidum* but not as dark green above, and has inconspicuous flowers in leaf axils. The flowers develop into cheese-shaped fruits, which are pale green to cream, with red seeds when ripe.

◄ The WEED Large-leaved Privet ***Ligustrum lucidum*** has opposite leaves, dark green above but paler underneath, and terminal sprays of small white flowers (here in bud) in summer, that form drooping bunches of small round purplish-black fleshy fruits in winter. *Ligustrum* is in the olive family, Oleaceae.

Similar shrubs

▲ The NATIVE ***Astrotricha floccosa*** (and the similar *Astrotricha latifolia*) in the Araliaceae family can be confused with Wild Tobacco because both have soft leaves thickly covered with hairs on their undersides. *Astrotricha*'s sprays of small white flowers ripening to small flattened fruits in pairs, and absence of stipule-like young leaves, are distinctive.

◄ The WEED Wild Tobacco Bush ***Solanum mauritianum*** has soft, hairy but more rounded leaves than *Astrotricha,* with extra tiny leaves where the leaf stalk joins the stem on older plants. Mauve flowers and cherry-sized round fruits that ripen to orange distinguish it.

Confusing creepers

▲ The NATIVE ***Commelina cyanea*** has blue flowers that distinguish it from the similar but weedy *Tradescantia. Commelina* leaves are generally less rounded, and more widely spaced. The plant grows strongly in spring and summer and dies back in winter.

▲ The WEED ***Tradescantia fluminensis*** (formerly *Tradescantia albiflora*) has white flowers, and never seems to stop growing, as implied in its common name, Wandering Jew. It grows very vigorously in wet shady places.

If you are not sure which plants are weeds in the bush, your local council may guide you to the nearest help in identifying them. As well as being bad for the bush, and possibly illegal, we'd hate you to waste your energy and enthusiasm pulling out the wrong plants!

Towards sustainability…

Aboriginal people were the first to live in and around Sydney, as we now know it. There were different clans and groups occupying specific areas, and moving around within those areas as food and resources changed with the seasons. People from different areas traded with each other. The bush provided fruits and berries, nuts, yams and bulbs, leaves and nectars. String was made from stem and root fibres, tools were carved from wood, containers were woven from strap-like leaves or made from trees, palm fronds and paperbark. Plant resources varied — close to the coast nectar was abundant when banksias flowered in autumn, for example, while on the Cumberland Plain plants with tubers and bulbs were plentiful.

▲ Aboriginal **grinding grooves** (foreground) remain on rocks within a few kilometres of the CBD, reminders of Sydney's former inhabitants. Stones, bones and shells were ground into sharp implements with the help of water from nearby rock pools or streams.

The Europeans who arrived in 1788 did not understand or recognise Aboriginal people's relationship with the landscape and how important this was in their social organisation. The Europeans' proprietary approach to land, and their introduction of European-style farming, displaced Aborigines from their traditional and necessary resources. The Europeans also brought smallpox with them, and this swept through the susceptible Aboriginal population as early as 1789.

Today there is belated recognition of the nature of Aboriginal ownership and custodianship of the land they occupied for so long. Descendants of Sydney's Aborigines of the 1780s are sharing their traditional knowledge with Australians of other backgrounds. As we walk through the bush today, we can recognise plants that provided Aborigines with food, tools and useful materials.

◄ The Grass Tree ***Xanthorrhoea resinifera*** provided Aborigines with multiple resources. Its name 'resinifera' means resin-bearer, and the yellow resin — here exuding from the fire-blackened trunk — was used to glue spear points in place, and mend canoes and wooden containers. *Xanthorrhoea* flowers provided nectar, the spikes were made into spear shafts, young leaf bases could be eaten and also provided a source for fire.

Changing attitudes

While the European settlement of Australia has been seen historically as a struggle against the land and the climate, of dogged settlers pitted stoically against the endless bush, our relationship with the bush has gradually changed as we have taken stock of the impacts we have wrought.

During pioneer times, despite the unceasing war waged against the trees to open up the land for farming, there was also an awareness of the natural beauty of the bush, and bushland flowers such as the Waratah *Telopea speciosissima* and Native Rose *Boronia serrulata* provided motifs for the decorative arts and architecture in 19th century Sydney.

That bushland had value in its own right, as a place of beauty and wildness, was recognised in suburban Sydney as early as 1879 by the establishment of Royal National Park. Ku-ring-gai Chase, Blue Mountains and Brisbane Water National Parks followed, establishing Sydney as a city of bushland.

Some native trees have been used in public park plantings for many years. Great Moreton Bay Figs *Ficus macrophylla*, native to the rainforests of the Illawarra south of Sydney, were planted over a century ago and form grand avenues in Centennial Park and the Sydney Domain. Brush Box, Silky Oak and Tallowwood from the north coast forests have also been used widely as Sydney street trees.

◀ Unfortunately there are still some residents who are prepared to commit terrorist acts rather than accept the trees as part of their view.

Taken for granted or taken seriously

Though bushland in Sydney suburbs has often been taken for granted, there has been increasing recognition since the 1960s that much is deteriorating as increased nutrient and sediment in runoff has promoted weed invasion and degradation.

Techniques of bush regeneration were established by volunteers to draw this situation to the attention of local authorities.

▲ Today the importance of **bush regeneration** is recognised and many trained professionals as well as local volunteers are employed in this activity. Dedicated volunteers and professionals remove many bags of weed material from bushland every year in an effort to stem the tide of weed invasion.

The widespread use of native plant species in gardens has become popular over the last couple of decades. Native plants can attract native birds and provide food for colourful insects such as butterflies. Native plants can be low maintenance and are environmentally friendly in using less water and less fertiliser than introduced ornamentals. It is becoming recognised that suburban gardens should complement adjacent bushland and that it is valuable to use local native species to improve the habitat for local wildlife as well as to minimise the threats posed by exotic species that seed into the bush.

We see our suburban areas as long-term living places. We see the natural values that the adjacent bushland parks, creeklines and bush remnants provide. We can also see the problems of weed growth and bushland deterioration. The challenges for urban bushland management are now to focus on true long-term ecological sustainability, to be serious about maintaining local populations of plants and wildlife for the truly long-term. Future generations have the right to see and experience Sydney bushland as we are able to see and to explore it.

OUT AND ABOUT IN SYDNEY'S BUSHLAND

So where is a good place to begin 'seeing' Sydney's bushland? There are many places, and there is probably somewhere not far from where you are. To give you a start we have chosen a selection of places that we like and which provide a range of bushland encounters and are relatively well signposted and easy to find. We have tried to choose places accessible by public transport — train, bus or ferry, though this has not always been possible as some well-known places are only accessible by car.

We have also tried to indicate places you might enjoy in a morning, a day or a little longer, though of course you may need longer for serious botanising, as we invariably do. There are many other places to visit — see visitor centres and local guidebooks for more ideas.

Places to visit **in a morning or half-day** ◖ are mostly around Sydney Harbour or within 10 km of the CBD. The Royal Botanic Gardens Sydney and Sydney Harbour National Park — South Head, Neilsen Park, Bradleys Head, North Head and Manly Scenic Walkway — are clustered around the Harbour and generally accessible by public transport — by bus or, more salubriously, by ferry. A little further away are Garigal National Park to the north and Lane Cove National Park to the north-west, Bicentennial Park and the Olympic site to the west, and to the south is Botany Bay National Park.

If you have **a day** ● we have suggested places in the magnificent coastal major national parks that fringe Sydney — Royal and Ku-ring-gai Chase National Parks. Or the relatively newly proclaimed Western Sydney reserves — Scheyville and Cattai National Parks, Castlereagh and Agnes Banks Nature Reserves and Mount Annan Botanic Garden. Royal and Ku-ring-gai are accessible by train though not for all the places we suggest.

If you are lucky to have **a bit longer** ●◖ we have suggested the renowned Blue Mountains west of Sydney — Glenbrook in the lower mountains, or Mount Tomah Botanic Garden and Katoomba for the upper Blue Mountains. Glenbrook and Katoomba are accessible by train but not Mount Tomah.

So grab your things and go! P.S. If you want a book to help with identifying plants see our reference list.

◀ Pink twisted branches of *Angophora costata* trees frame water views along this inviting track on Bradleys Head in Sydney Harbour National Park. Photo: Jaime Plaza

A tranquil setting beside the Opera House and the Harbour

Why not begin with a morning at the Royal Botanic Gardens Sydney, in its tranquil setting near the Opera House, and the harbour shoreline of Farm Cove. It was here in January 1788, beside the small creek that runs down into the cove, that European settlers first cleared land for agriculture.

▲ The site of **Australia's first farm** has been planted with examples of the crops grown in the early days of European settlement in Sydney.

To the Cadigal people, the Aboriginal inhabitants prior to 1788, Farm Cove was known as *Woggan-ma-gule*. It was part of a wider country they occupied, *Cadi*, stretching from present-day Darling Harbour to South Head. The land now occupied by the Royal Botanic Gardens can be recognised as a 'first frontier' between Aboriginal and European societies.

► The indigenous history of the site is now recognised in a display between Botanic Gardens Creek and the 'First Farm', entitled ***Cadi Jam Ora — First Encounters.*** The garden features plants that originally grew here and were used by Aborigines for food, shelter, tools, medicines, art and ritual. Interpretive stories and events focus on the meeting of two vastly different cultures in the first years of European colonisation. *Cadi Jam Ora* means 'I am in Cadi'.

The bushland of *Woggan-ma-gule* continued to disappear though farming here was abandoned when better soils for agriculture were found elsewhere. The site became a Government Garden and then a botanic garden.

In the nineteenth century the Botanic Gardens developed along traditional European garden and landscaping lines — mixing foreign or exotic species introduced for potential horticultural or ornamental use, with occasional plantings of notable Australian species. These include native rainforest trees and prominent architectural species, such as Hoop Pine *Araucaria cunninghamii* and Norfolk Island Pine *Araucaria heterophylla*.

Over the years natural features of the landscape were gradually destroyed — low-lying land was filled and smooth sandstone sea walls replaced the natural shorelines. Formal gardens, informal lawns, curving paths and special plantings such as the Palm Grove now blend in a 'picturesque landscape'.

Australian plants are scattered throughout the Gardens, but feature in the Solander Garden, the rainforest area, the Rockery near the Opera House, and along the eastern fence north of the Henry Lawson gate. Botanical displays featuring plants from other parts of the world include the Sydney Tropical Centre, the Herb Garden, the Rose Garden, the Succulent Garden and the Palm Grove.

▲ The building shown here was once the Museum and later the Library for the National Herbarium of New South Wales. The National Herbarium Collection, the meticulously curated collection of dried specimens of NSW and Australian plant species, is the heart of the resources used for botanical research by staff of Plant Sciences Branch. It is now housed in the new Robert Brown Building immediately adjacent. If you are short of time, why not take a guided tour of the Gardens by trackless train.

Joseph Banks and Daniel Solander collected Sydney's first botanical specimens at Botany Bay in 1770. Live plants of some of these species are featured in the Solander Garden near the original Herbarium building.

▲ Nearby is the **Rare and Threatened Plants Garden**. Signs here tell of the plight of so many of our plant species, not only Australian, but also species from elsewhere in the world. Botanic Gardens can provide protection for some of these species. Close by is a plant of the recently discovered Wollemi Pine *Wollemia nobilis.*

At the **Visitor Centre,** a short stroll away, a wide range of botanical, gardening and tourist books is available.

▼ The Fernery, near the Palm Grove, displays ferns from both Australia and other parts of the world. Here you can see ferns in a variety of shapes and sizes. Sydney ferns include *Dicksonia antarctica* and *Blechnum nudum.*

Venerable old timers

▲ Beside the historic Macquarie Wall, originally built to separate the 'Government Gardens' from the swampy foreshore, are perhaps the first formal plantings of an Australian native species. Here an avenue of Swamp Mahogany trees ***Eucalyptus robusta*** was planted by Governor Lachlan Macquarie in 1816. Some of these trees are still thriving after nearly 200 years. Swamp Mahogany trees grew naturally in swamp forest around Sydney and it is likely that the seed was collected from local trees nearby. As a winter flowering tree it is an important provider of nectar for wildlife when few other flowers are available.

▲ There are also a few remnant native trees growing in the Gardens — surviving from a time when there was still bushland here. Three Swamp Oaks ***Casuarina glauca*** still grow near the Maiden Pavilion on a slope that was part of the original Farm Cove foreshore before its mudflats were filled in the 1860s to become the Lower Gardens. Remnant Forest Red Gums *Eucalyptus tereticornis* and Blackbutts *Eucalyptus pilularis* may also be found in the Gardens and Domain.

The Sydney Domain

▲ The parkland of the Domain borders the Botanic Gardens. From the Botanic Gardens, walk out to Mrs Macquaries Point, named after Governor Macquarie's wife. A loop road was constructed here in 1816 and **Mrs Macquaries Chair**, cut into the Sydney sandstone and now shaded by Moreton Bay Fig trees, provides a panoramic view of Sydney Harbour. Nearby views of the Opera House and Sydney Harbour are world famous.

Walking in the Domain, you may pass Forest Red Gums and Blackbutts that are descendants of the original forest trees. There are also plantings of the magnificent Moreton Bay Fig *Ficus macrophylla*, a native of the rainforests of the Illawarra region south of Sydney.

▲ The local Port Jackson Fig ***Ficus rubiginosa*** is also prominent. It grows naturally clinging to cliff faces, overlooking the Harbour and other Sydney waterways. Here its branches frame one of Sydney's most-photographed views.

► Bushland survived on Mrs Macquaries Point until the 1920s, when it was removed to create the open grassy park-like landscape you see today. A recent program to re-establish native woodland vegetation in the Domain along the western foreshores of Woolloomooloo Bay has been very successful. Here plant material collected from local sites has been planted amongst the natural sandstone rock outcrops as part of **Mrs Macquarie's Bushland Walk** — displaying the sort of vegetation that Mrs Macquarie may have seen as she drove out to the point in the early 1800s.

You may see bushland that is still intact by looking across the Harbour from Mrs Macquaries Point to Bradleys Head and Mosman. Bushland survived here as part of a military reserve. Now it is part of Sydney Harbour National Park, and you can catch a Taronga Zoo ferry to visit it.

▼ Forest Red Gums *Eucalyptus tereticornis*, which grew in the bushland that once covered harbour headlands, overlook two of Sydney's most famous icons.

Scenic vistas of the Harbour

Spectacular headlands, scenic vistas of the Harbour, small sandy beaches and delightful picnic spots make **Sydney Harbour National Park** a great place to begin to get to know Sydney's bushland plants. The Park includes most of the beautiful bushland fringing Sydney Harbour including North and South Head, Bradleys Head and Middle Head, as well as some of the Harbour islands. One popular place is The Gap at South Head.

The Gap and South Head

This destination is high on any tourist's visiting list. Take the Watsons Bay bus, or a ferry from Circular Quay. Once a sleepy fishing village, Watsons Bay is now popular for its seafood as well as its scenery.

▲ To the west you look back, between wooded headlands and leafy suburbs, to the city towers. The Coast Banksia trees ***Banksia integrifolia*** in the foreground are the largest most abundant bushland plants in this coastal sandstone scrub.

The coast between The Gap and South Head is shared by Sydney Harbour National Park and the naval establishment HMAS Watson.

▲ Walk up to the cliff-edge above The Gap for some spectacular views. Cliff walls of Hawkesbury Sandstone tower above the breaking waves of the Pacific Ocean.

When you can take your eyes off the spectacular scenery, you will see plants of Sydney's coastal sandstone heath and scrub around you. These are hardy plants, surviving here despite a long history of the area's military use.

It is perhaps surprising that any bushland at all remains on the Harbour foreshores of a large city with rapacious demands for waterfront urban living. Much of today's harbourside bushland only survived because it was within lands originally reserved for military defences. When no longer needed for defence there was strong public support for these lands to become national park. You are likely to come across ruined and derelict fortifications in many of the bushland places you visit.

▲ Shrubs beside the cliff top walk are low and wind-pruned. They also tolerate salt in wind and sea spray. Small round yellow clusters of ***Melaleuca nodosa*** flowers can be seen here in October. You may also be lucky enough to see a hawk hovering above the cliff top in search of prey.

◄ Among the sculptured rocky outcrops and the Coast Banksia trees *Banksia integrifolia* you may see smaller shrubs like the felty-leaved daisy ***Olearia tomentosa***. Daisies belong to the Asteraceae, formerly Compositae, family. This *Olearia* flowers profusely in October.

▲ ***Westringia fruticosa*** is in the mint family Lamiaceae, formerly Labiatae. *Westringia* is a common shrub on coastal headlands, particularly those facing the ocean. Its ability to withstand the salt spray has made it a popular garden plant in coastal suburbs.

▲ Prickly ***Hakea teretifolia***, with its needle-like leaves, is best avoided by bare-legged hikers, but in November its cream flowers can be admired.

Along the cappuccino coast…

Nielsen Park

The largest bushland area on the Harbour's southern foreshore is **Nielsen Park, Vaucluse**, part of Sydney Harbour National Park. It is accessible by bus from the City.

◀ From nearby headlands there are splendid views towards Manly and the city, here framed by a Coast Banksia tree, ***Banksia integrifolia.*** Despite the small amount of bushland, a rare new sheoak species, *Allocasuarina portuensis*, was found here as recently as the 1980s, and has been found nowhere else.

▲ Coast Banksias ***Banksia integrifolia*** are common bushland plants on the headlands here. In September they have beautiful yellow flowers. These develop into woody cones. Unlike other Sydney banksias *Banksia integrifolia* releases its seed every year, without needing fire or drought to open the seed capsules.

▲ Shark Beach at Nielsen Park is a popular swimming and picnicking spot and nearby is just the place to have a cappuccino and begin a relationship with the Harbour's foreshore bushland.

▲ The **Hermitage Foreshore Walking Track** leads through bushland from the western end of Shark Beach. It follows along the sandstone harbour foreshores, past Strickland House, for about 1.4 km, providing spectacular views of the Sydney city skyline and northern Sydney.

Although the strip of bushland below the houses is often very narrow, there are still places along the track where the sandstone woodland and heath are wide enough to let us imagine how the Harbour and its bushland shores looked before the city was ever thought of.

Along here you can see trees of Coast Banksia *Banksia integrifolia*, Tuckeroo *Cupaniopsis anacardioides*, Cheese Tree *Glochidion ferdinandi* and Bangalay *Eucalyptus botryoides*. There are extensive groves of the shrubby *Kunzea ambigua* on rock platforms together with occasional *Monotoca elliptica* shrubs. In one place, near the convent cemetery, Coastal Teatrees *Leptospermum laevigatum*, have grown old into huge, twisted shapes.

◀ The native Port Jackson Fig ***Ficus rubiginosa*** twists and clings to vertical rock faces. Its small fruits would have formed part of the Aboriginal diet.

In August–September the small shrubs of the sandstone woodland and heath are flowering — deep pink *Crowea saligna*, yellow *Dillwynia retorta*, red and white *Epacris longiflora*, pale pink *Philotheca buxifolia*, and rusty *Lasiopetalum ferrugineum* along with the beautiful *Pandorea pandorana* vine. One might be lucky enough to see a 'grove' of the tiny ground orchid *Pterostylis nutans*, the Nodding Greenhood.

▲ Unfortunately, nearby housing has changed conditions for the bush along this track, and it is now a mix of native plants and exotic weeds. Many of the weeds have originally naturalised from gardens, particularly in narrow disturbed sites and downslope from houses where water and nutrients concentrate. Spread of weeds with fleshy fruits, such as *Lantana camara* and Sprengeri Fern *Protasparagus aethiopicus*, is further enhanced by fruit-eating birds, both native and introduced, which distribute the seeds widely. The absence of fire from bushland remnants for long periods restricts the recruitment of native species and also favours weeds.

To Bradleys Head and beyond … by ferry

Bradleys Head

Why not explore Sydney Harbour's bushland by ferry? **Bradleys Head**, named after Lieutenant William Bradley, of the First Fleet flagship H.M.S. *Sirius*, is a short ferry ride across the Harbour from Circular Quay on the Taronga Zoo Ferry. It is part of Sydney Harbour National Park.

▲ From the Zoo wharf a path follows the shoreline south-east along the side of the hill to the sandstone headland, passing woodland trees of Bangalay *Eucalyptus botryoides* and the pink twisted limbs of Smooth-barked Apple *Angophora costata*.

▲ At the southern end are old gun emplacements and fortifications, built in fear of Russian threats in the 1860s, though the bush around has long since regrown. Like many bushland areas in Sydney Harbour National Park, Bradleys Head was once reserved for defence purposes. This saved these areas from housing development and subsequently they became public parkland.

▲ Bush plants at Bradleys Head include the shrub ***Dodonaea triquetra*** shown here, as well as *Banksia serrata, Glochidion ferdinandi, Pandorea pandorana, Pittosporum revolutum* and *Pittosporum undulatum*.

In places shrubby weed species predominate. Following conflict over the use of controlled burning in the 1970s, bush management principles were developed in this reserve by Eileen and Joan Bradley who advocated the selective removal of exotic species rather than bulldozing and unselective spraying.

▲ Along the tracks there are many native species to see in a beautiful harbour setting.

From Bradleys Head a path leads around to bushland on Chowder Head and Clifton Gardens. Further round is bushland on Middle Head. To the west of the Zoo Wharf is Little Sirius Cove, where in the 1890s Australian landscape painters, including Tom Roberts and Arthur Streeton, were inspired by the Harbour's bushland setting.

North Head

If you have more time take the big Manly Ferry to Manly, a marvellous 30 minute cruise with plenty of opportunity to contemplate the beauties of the Harbour environment. From Manly take a taxi or perhaps walk out (about 4 km) onto the spectacular sandstone massif that is **North Head**. On the way you will pass through woodland on the sheltered west facing slopes. The historic Quarantine Station occupies much of this area.

▲ On the top of the headland, heath and scrub extend across to the cliffs fronting the Pacific Ocean. The **Fairfax Walking Track** provides a short circular walk with interpretive signs. Common plants here are *Banksia ericifolia, Allocasuarina distyla, Angophora hispida, Baeckea imbricata, Acacia suaveolens, Kunzea ambigua, Leptospermum laevigatum, Melaleuca armillaris, Epacris longiflora, Leucopogon microphyllus, Grevillea speciosa, Darwinia fascicularis* and *Hakea teretifolia.*

Periodic fire is an important part of the ecology of coastal heath and you might see patches of vegetation at different stages of growth after fire.

▲ The views back up the Harbour from North Head are unforgettable.

◀ *Melaleuca armillaris* is found naturally in heath on coastal headlands near Sydney. However it has been planted more widely, particularly along motorways and in suburban landscaping

A shorter walk is from the ferry wharf, along The Corso to the surf beach at Manly and then around the foreshore to **Shelly Beach**. Cabbage Tree Bay here was named for the Cabbage Palms *Livistona australis* that once grew in a patch of rainforest. Although these have gone, there are some planted specimens nearby. There are still rainforest vines growing here, including *Cissus hypoglauca*, as well as Coast Banksia *Banksia integrifolia*, Cheese Trees *Glochidion ferdinandi*, Sweet Pittosporum *Pittosporum undulatum* and Lilly Pillys *Acmena smithii*. Along the top of the cliffs are shrubs of *Melaleuca nodosa*.

Spit to Manly!

Manly Scenic Walkway

Manly and The Spit are connected by the Manly Scenic Walkway. The Spit Bridge crosses Middle Harbour to link the narrow sandy peninsula of The Spit with the steep sandstone hillside of Seaforth. The walkway track begins at the Spit Bridge, runs along Clontarf Beach — there is a detour to Grotto Point — past Crater Cove, Dobroyd Head, Reef Beach, Forty Baskets Beach, and around the harbour foreshores of North Harbour, to Manly. To start the walk you can catch a bus to The Spit, then return to Circular Quay on the Manly Ferry, or, of course, you can walk in the other direction.

▲ The walkway is clearly marked and passes for much of the way through **heath** and woodland. Both visitors and local people use the track and enjoy its diverse range of colourful shrubs. Common heathland shrubs are *Banksia ericifolia, Allocasuarina distyla, Hakea teretifolia* and *Kunzea ambigua.* You may be lucky enough to see King Parrots in the woodland, or a flock of Yellow-tailed Black Cocktoos on Dobroyd Head. White Flannel Flowers, *Actinotus helianthi*, may catch the eye.

▼ There are magnificent views over the Harbour from Dobroyd Head looking across the low shrubby heath of this high sandstone headland.

▲ The bright red 'bottlebrushes' of ***Callistemon linearis*** are made up of many individual flowers. Together they make a colourful show to attract birds as pollinators.

▲ Despite proximity to the city these areas are occasionally burnt. Bushland along the Manly Scenic Walkway was severely burnt by a hot fire in November 1990.

▲ After a fire, ***Isopogon anemonifolius***, known as Drumsticks, is one of the Proteaceae family shrubs that resprouts.

▲ Bushland is not destroyed by a single fire. It's hard to find traces of the 1990 fire in the regrown bush in 1999. Indeed fire is necessary to maintain many of the plant species. In the foreground are white Flannel Flowers ***Actinotus helianthi*** which are killed by fire but regrow from soil-stored seed that germinates afterwards.

▲ It is hard to imagine that periodic bushfire is an important natural element in the ecology of the sandstone woodland — such as the woodland here behind peaceful Reef Beach.

A little further up the Harbour — mangroves and saltmarsh

'Those Coves above where the ships lay were surrounded by mangroves and had Mud flats at the bottom, those below had sandy beaches most of them.' Recorded by Lieutenant William Bradley of the First Fleet flagship *Sirius*, who was involved in the first survey of the Harbour in February 1788. We can still see some of the sandy beaches in the lower Harbour, but much of the original mudflats, mangroves and saltmarsh have been filled in.

Bicentennial Park

Bicentennial Park at Homebush Bay, about 12 km upstream along the Parramatta River from the CBD, was established in 1988 to mark 200 years since the European occupation of Sydney. Much of the open space was formerly saltmarsh, mangrove and mudflats that were progressively filled between 1890 and 1970. Some areas of mangroves survived however and you can walk through them on boardwalks.

▲ Grey Mangroves ***Avicennia marina*** form a low forest above a carpet of their seedlings. The boardwalk provides access and keeps feet dry when the mangroves are flooded by high tides.

There is an **Information Centre,** a field studies centre, a bird hide, plenty of picnic areas and a more recently constructed freshwater lake, Lake Belvedere. With both saltwater and freshwater habitats Bicentennial Park is also a place to see many different types of birds.

Olympic site

The Olympic site is adjacent to Bicentennial Park and includes remnants of saltmarsh and mangrove as well as forest that survived within the exclusion zone around the former Newington Armaments Depot.

◄ ***Wilsonia backhousei*** is a tiny saltmarsh plant in the Convolvulaceae or Morning Glory family that has survived on the edge of Haslams Creek. Salt tolerant species provide a link with the salt lakes of inland Australia and some species of *Wilsonia* grow in both coastal saltmarsh and inland salt lakes.

◄ Scribbly Gums, Turpentines and Stringybarks at Newington in a surviving patch of bushland, rare in this part of Sydney. It is part of the shale **Sydney Turpentine Ironbark forest**, though the Scribbly Gums indicate a sandstone influence. Because of their vulnerability to disturbance, access to these areas is restricted.

'Bungaroo', a First Fleet connection

Garigal National Park in northern Sydney includes sandstone slopes of the Middle Harbour Creek valley and other creeks draining into Middle Harbour, the northern arm of Sydney Harbour. Other parts of the park form part of the catchment of Deep Creek, which flows into Narrabeen Lagoon.

There are many tracks though the different parts of the Park, with access from adjacent suburbs of St Ives, Davidson, Forestville, Frenchs Forest, Allambie Heights and Killarney Heights.

'**Bungaroo**' is the site on Middle Harbour Creek where Governor Phillip and officers from the First Fleet camped in April 1788, when first exploring the area. It is still in natural condition today, and it is possible to feel as though you are in bushland many miles from anywhere, despite the proximity of suburbia on the surrounding ridges.

The track begins from Hunter Street, St Ives at a log sign 'Founder's Way'. Fit people can walk down to the creek and back in an hour, but allow more time for looking at the plants and enjoying the views.

▲ Scribbly Gums ***Eucalyptus haemastoma*** are abundant in the ridgetop woodland at the beginning of the track, the thin sandy soil here and on the hillside ledges supports only widely-spaced, low-growing trees. Woodland species here include *Banksia serrata, Dillwynia floribunda, Grevillea buxifolia* and *Pultenaea stipularis.*

Shrubby heath grows where the soil is poorly drained. Characteristic species *Banksia ericifolia, Allocasuarina distyla*, prickly *Hakea teretifolia* and occasional *Angophora hispida* may be seen.

At the edge of the ridge there are mallee eucalypts — trees with multiple thin trunks arising from the ground instead of a single thick trunk. These are *Eucalyptus sieberi* growing in mallee form. The track traverses a gently sloping ledge, then begins to drop more sharply to the creek, with *Eucalyptus piperita* and *Angophora costata* in open-forest. Here *Eucalyptus sieberi* grows as a single-trunked tree in the less harsh conditions.

◀ As you descend the rocky slope, you may notice bright pink flowers of ***Crowea saligna***; these shrubs seem to flower sporadically through the year, unlike the local *Boronia* species, also in the Rutaceae family, that have regular annual flowering times.

◀ Down at the creek, large 'stepping stone' rocks mark the end of navigation. The tidal limit is a bit further upstream. Swamp Oaks *Casuarina glauca* indicate brackish conditions here.

Upstream, characteristic riparian scrub species fringe the creek beside sandy terraces where Blackbutts *Eucalyptus pilularis* tower above a carpet of *Sticherus flabellatus* and other ferns.

Somewhere here, near the tidal limit, the First Fleet party camped on 16 April 1788. 'Here, in the most desert, wild, and solitary seclusion that the imagination can form any idea of, we took up our abode for the night,' and 'washed our shirts and stockings,' wrote Surgeon General John White.

A mid North Shore 'pleasure ground'

Lane Cove National Park, straddling the upper Lane Cove River on Sydney's mid North Shore, has had a chequered history of use — from early colonial timber yard, to farms and orchards, picnic and pleasure grounds, to today's National Park catering for both recreation and nature conservation. Access by car is from Fullers Bridge or Delhi Road on the Chatswood West side, or Lane Cove Road on the Macquarie Park side of the National Park.

Walking tracks run from the **Information Centre** near Fullers Bridge. There is also a small exhibit here with native animals. Tracks run westward along the northern side of the Lane Cove River to De Burghs Bridge, and along the southern side of the river. They connect with the Great North Walk — a walking track linking Sydney and Newcastle.

▲ Bushland in **Lane Cove National Park** has survived on the sandstone hillslopes of the main river valley, on sandy soils too steep for housing development, and too infertile for farms.

Years ago the riverflats with their fertile alluvial soils, were cleared of their native bush for orchards and small farms. These riverflats have since become grassy picnic sites and car parks. The level sandstone ridgetops with their more fertile shale soil cappings were gradually developed as suburbs, sporting ovals, light industry and cemeteries.

◀ ***Leptospermum polygalifolium*** here is a common tea tree along sandstone creeks around Sydney. A much rarer species, *Leptospermum deanei*, restricted to the upper Lane Cove River, upper Middle Harbour Creek, Marramarra Creek and Berowra Creek, was only recently discovered and described in 1989. Scientific knowledge of much of our bushland is still very limited.

◀ ***Hardenbergia violacea,*** False Sarsaparilla, is one of the earliest contributors to the spring wildflower display. Its intense purple flowers contrast with the yellow pea flowers.

◀ ***Lomandra gracilis*** is one of several *Lomandra* species that look like tufts of grass at first sight. In fact, *Lomandra* was previously included in the 'grass tree' family, Xanthorrhoeaceae, but now has its own family, Lomandraceae. The sprays of tiny yellow flowers and firm leaves are characteristic of *Lomandra.*

◄ Black Sheoak ***Allocasuarina littoralis*** is a small tree of the ridges and upper slopes. The pine-like foliage of this and other members of the Casuarinaceae family are really stems (termed 'cladodes'), with leaves reduced to whorls of scales. In *Allocasuarina littoralis* male and female flowers are produced on separate plants — the showy orange colouring in August–September being due to the male flowers.

▲ On the sandstone hillsides, between rocky outcrops and twisted gum tree trunks, colourful wildflowers can be seen depending on the season. The yellow pea flowers of the shrub ***Pultenaea stipularis*** are conspicuous in August–September.

◄ The light green fronds of the soft fern ***Calochlaena dubia,*** foreground, grow in the shelter of lower hillslopes, especially where there is some moisture seepage. Also here a Smooth-barked Apple *Angophora costata*, having shed its outer bark since the last fire, reveals its smooth pink trunk. This bark contrasts with the charred persistent fibrous bark of the nearby Blackbutt *Eucalyptus pilularis.*

Large trees, mostly Blackbutts *Eucalyptus pilularis*, and a few Sydney Blue Gums *Eucalyptus saligna*, have regrown on the alluvial flats since the timber-getting days.

◄ Turpentines ***Syncarpia glomulifera*** may also be prominent. These are in the Myrtaceae family, the same family as the eucalypts, but have distinctive fruit capsules and leaves wth grey undersurfaces.

Here, in the upper Lane Cove River valley, the relationships between the natural landscape and a history of increasingly intensive urban land use is evident. Enriched by increased nutrients in runoff water from developed ridgetop areas, the riverbanks and side creeks have been overgrown by tangles of weeds. There are active bush regeneration programs in the Lane Cove National Park and in other bushland in the area, including weeding and replanting along the riverbanks — so you may see signs asking you to avoid regeneration areas. Bush regeneration is an important aspect of bushland management in reserves adjacent to suburban areas and often local people are enthusiastically involved.

Bound for Botany Bay

South of Sydney Harbour lies the broad expanse of Botany Bay, known to the Aboriginal people as *Kamay.* Here, in April 1770, the naturalists aboard Captain James Cook's *Endeavour*, Joseph Banks and Daniel Solander, encountered an unusually large number of previously unknown plants. Cook changed the name from Stingray Bay to Botany Bay, in recognition of this new and impressive flora. Principally as a result of Banks' enthusiasm, Botany Bay, in Cook's 'New South Wales', was subsequently chosen by the British Government as the site for a penal settlement.

Captain Cooks Landing Place, Kurnell

◄ The historic site of **Captain Cooks Landing Place** on the southern shore of Botany Bay is now marked with monuments and plaques. Trees of Bangalay ***Eucalyptus botryoides*** were common in the woodland along the foreshores of Botany Bay here.

▲ However the real thrill for the botanical explorer is further round to the east, in the woodland and **heath** that survives on the clifftops of Cape Solander. Here you can still see the rich and diverse flora that so impressed Banks.

The **Cape Baily Coastal Walk** runs south from Cape Solander along the coast to Cape Baily lighthouse through extensive areas of heath. Common shrubs are *Banksia ericifolia,* named after Banks, and *Darwinia fascicularis*, named after fellow scientist Erasmus Darwin, grandfather of naturalist Charles Darwin.

▲ ***Dampiera stricta*** in the Goodeniaceae family is named after William Dampier who collected specimens of Australian plants in Western Australia as early as 1699.

La Perouse and Cape Banks

▲ On the northern headland of Botany Bay, small areas of **coastal sandstone heath** and woodland, also part of Botany Bay National Park, can be seen closer to the city at **La Perouse**, on the headlands from Cape Banks to Henry Head and around the foreshores to Congwong Beach. Plants along here include *Banksia serrata, Coymbia gummifera, Xylomelum pyriforme* and *Actinotus helianthi.*

Access to these areas is by track from Cann Park (opposite bus terminus and near car parking area) near Congwong Beach.

Different **coastal dune heath** on deep sand occurs at nearby Jennifer Street (off Anzac Parade). This is a small vestige of the **Eastern Suburbs Banksia Scrub** that once covered most of Sydney's Eastern Suburbs and which is now listed as an Endangered Ecological Community under the *NSW Threatened Species Conservation Act*. From the air — a view many Sydney travellers will experience given the proximity of the airport — there are large areas of green visible among the suburbs. But these are golf courses and playing fields, not the native Eastern Suburbs Banksia Scrub that once grew here. Only a few small remnants of this former vegetation survive.

▲ Signs with map and information, mark the beginning of a boardwalk through the **Eastern Suburbs Banksia Scrub** at Jennifer Street.

▲ The boardwalk protects this small patch from excessive trampling. Shrubs along here include ***Acacia longifolia*** and the small-leaved ***Kunzea ambigua*** which has small starry white flowers, heavy with the scent of nectar between October and January.

▲ The yellow flower spikes of Sydney Golden Wattle ***Acacia longifolia*** can be seen in July and August.

◄ A very conspicuous plant is the Grasstree ***Xanthorrhoea resinifera***, with its long thin spiky leaves fanning out from the ground and shimmering in the lightest breeze. Tiny white flowers clustered along a woody 'spear' attract honeyeaters. Older plants may have a 'skirt' of old leaves hiding their short stocky rough trunks.

Other shrubs along here are the Coastal Teatree *Leptospermum laevigatum* with white flowers like tiny single roses in September, and the Wallum Banksia *Banksia aemula*, with its thick serrated leaves, warty bark, and large greenish flower heads between March and June. The *Banksia* and *Xanthorrhoea* are characteristic of Eastern Suburbs Banksia Scrub.

◄ The tiny white flowers of ***Monotoca elliptica***, a shrub in the Epacridaceae or southern heath family, are seen in spring.

Visit the oldest national park in Australia — a place of great botanical diversity

Royal National Park is situated about 30 km south of Sydney past Sutherland. The northern half of the park is a broad Hawkesbury Sandstone plateau dissected by the Hacking River and its tributary creeks. Here heath and woodland predominate. Moist eucalypt forest predominates further south where underlying layers of shale-rich Narrabeen Group rock are exposed in deep sheltered valleys. You can see the heath vegetation at **Wattamolla** though it occurs at many other places. The **Forest Path Walking Trail** at Bola Creek is one of the best places in the Sydney area to experience rainforest and moist tall eucalypt forest. There is a **Visitor Information Centre** at Audley.

▲ Cliff-top plants include *Allocasuarina distyla,* white-flowered *Westringia fruticosa* and *Philotheca buxifolia. Baeckea imbricata* shrubs are knee high and lower down the cliff sides in sheltered places facing into bays are tussocky clumps of *Lomandra longifolia.*

Wattamolla

▲ **Wattamolla** is a small beach and lagoon where the explorers George Bass and Matthew Flinders landed the Tom Thumb in 1796 — and where they are supposed to have entertained the local Aborigines by cutting their hair. Apart from the carpark and grassy picnic area, the surroundings are much as they would have been 200 years ago.

There are spectacular views of sandstone cliffs with wind-pruned heath clinging to the headlands at nearby Boy Martin Point (Boy William Martin accompanied Bass and Flinders), a short walk through heath and low scrub along the southern side of the inlet.

Nearby, twisted trunks of *Melaleuca armillaris* shrubs lean away from the wind — in October, cream bottlebrush flowers are sprinkled over their dark green crowns. These attractive shrubs have been planted in gardens and landscaping all over Sydney, but cliff edges are their natural habitat.

▲ *Senecio lautus* **subspecies** *maritimus* is a small native daisy confined to crevices on the coastal sandstone cliffs.

From the Coastal Walk, which leads north from here to Bundeena and south to Garie Beach, are spectacular views of the coastline and coastal heath. A ferry runs from Bundeena to Cronulla railway station.

▲ Away from the salt spray there's a bigger variety of small shrubs, giving particularly colourful wildflower displays in August and September. Here you may see *Banksia serrata, Allocasuarina distyla, Banksia ericifolia, Dillwynia floribunda, Dillwynia retorta, Grevillea oleoides, Aotus ericoides, Boronia ledifolia, Darwinia fascicularis, Xanthorrhoea resinifera, Ricinocarpos pinifolius, Leptospermum squarrosum.* The shrubby heath is interspersed with groves of woodland in sheltered places.

◄ The red and white tubular flowers of ***Epacris longiflora*** are sometimes conspicuous on moist rock ledges and give it the common name of Native Fuchsia or Fuchsia Heath.

◄ Where thin lenses of shale have formed poorly drained 'benches' on the sandstone, alternating moist and dry conditions predominate. White-flowered ***Epacris obtusifolia*** and ***Epacris microphylla***, and pale pink-flowered ***Sprengelia incarnata*** mark these areas in early spring. All three have the characteristic small hard leaves of the Epacridaceae or 'southern heath' family.

Mallee eucalypts, with multiple trunks growing to about four metres tall, are also characteristic of these windswept coastal heaths; the Port Jackson Mallee *Eucalyptus obstans* is one species that is locally common here.

The best rainforest in Sydney!

Bola Creek

Vegetationally speaking, the antithesis of low windswept heath is tall moist rainforest. **Forest Path Walking Trail** at Bola Creek provides the opportunity to see some of the best rainforest in the Sydney area. From the southern entrance gate to Lady Carrington Walk, walk to Bola Creek. The creek banks are lined by rainforest. Here the **Forest Path** leads you into the darkness beneath a dense canopy of rainforest trees.

▲ Amongst the variety of trees forming the **rainforest** canopy, most common are Sassafras *Doryphora sassafras* with fragrant coarsely toothed leaves, Coachwoods *Ceratopetalum apetalum* with smooth grey trunks characterised by attractive patterns of lichens, and *Schizomeria ovata* trees, known as Crabapples. These trees belong to two families that contain predominantly rainforest species — Monimiaceae and Cunoniaceae.

Lilly Pilly trees *Acmena smithii,* with small mauve fruits attractive to Currawongs and other fruit-eating birds, are also common along the banks of Bola Creek and the Hacking River. This is one of the soft-leaved or mesic species in the Myrtaceae family, which also includes the eucalypts.

Rainforest tree trunks and branches are often decorated with epiphytes — 'perching' plants — *Asplenium australasicum* Bird's Nest Fern and *Platycerium bifurcatum* Elkhorn, are two striking examples along the Forest Path.

◄ Rainforest trees can be hard to identify when their canopies are high and leaves out of reach. Easily distinguished amongst them here are the Cabbage Palms ***Livistona australis,*** with their large circular fan-like fronds.

◄ Lianes trailing over ground plants and up tree trunks include the Twining Guinea Flower ***Hibbertia dentata***, which has bright yellow flowers the size of a 20-cent piece. Other trailing plants are *Marsdenia rostrata, Sarcopetalum harveyanum*, and the red-fruited *Stephania japonica.*

Ground ferns here amongst the young rainforest saplings include Giant Maidenhair *Adiantum formosum*, *Doodia aspera*, a Rasp Fern with leaves that feel like sandpaper, species of Shield Fern *Lastreopsis*, Sickle Fern *Pellaea falcata*, and occasional clumps of Gristle Fern *Blechnum cartilagineum* with large fishbone-like fronds.

There are also tussocks of *Gahnia melanocarpa*, one of the Saw-sedges with sharp leaf margins that will cut your fingers, *Gymnostachys anceps* known as Settler's Flax because its strap-like leaves are not sharp-edged and were used for binding by early settlers and the grass *Austrostipa verticillata* that clumps like miniature bamboo.

On the way to the rainforest you will have passed through tall open-forest with majestic trees up to 50 metres high, and maybe two to three centuries old. The trees with pale smooth trunks are *Eucalyptus botryoides/saligna*, a variant of Sydney Blue Gum, while those with thick brown fibrous bark are Turpentines *Syncarpia glomulifera*, closely related to the eucalypts. In October, Turpentine canopies are covered in cream flowers, but their gun-turret fruits can almost always be found littering the ground.

Beneath these forest giants are smaller trees with green branches and feathery leaves — *Acacia parramattensis,* the Parramatta Wattle. A scrambling shrub with oval leaves and yellow flowers, *Goodenia ovata*, is common here, as are the ferns mentioned above for the rainforest, the tussocky grass *Poa labillardierei* and also Bracken fern, *Pteridium esculentum*. One of the interesting forest understorey shrubs of more open areas is *Pimelea ligustrina* which has strong bark likely to have been used as twine by Aborigines. Unfortunately, in the part of the forest Path along the Hacking River a few weeds have also invaded these more open areas.

The Forest Path continues beyond the rainforest, which is confined to the Bola Creek valley, and along the Hacking River through moist tall open-forest with extensive stands of Cabbage Palms. It is well worth the 4.5 km walk.

Heathcote to Waterfall

Other interesting parts of Royal National Park are accessible from Heathcote and Waterfall railway stations. Walking tracks link these stations and radiate to other parts of the Park passing through extensive heath, sedgeland and woodland.

◄ The Waratah ***Telopea speciocissima*** in the Proteaceae family, used as a state emblem for New South Wales, occurs in the woodland and forest habitats in Royal National Park.

▲ ***Doryanthes excelsa***, the Gymea Lily, is a prominent plant in woodland on sandstone in Royal National Park and is often found among the pink-trunked Smooth-barked Apples *Angophora costata*. Together with the NSW north coast species *Doryanthes palmeri*, these two species make up the family Doryanthaceae. Because of their colour and size they are sometimes mistaken for waratahs at a distance.

◄ A cup of coffee at Audley on the Hacking River is a welcome beginning or ending to a day's bushland exploring in Royal National Park.

Another great coastal National Park

Ku-ring-gai Chase National Park is Royal's northern counterpart. Established in 1894, it includes extensive sandstone plateaus fringed by the creeks and bays of the lower Hawkesbury River and Broken Bay. The landscape here is perhaps less dramatic than the windswept heaths and moors of Royal, but there are quiet bays and coves with sandy beaches and wooded headlands, extensive ridges with heath and woodland, and sheltered valleys with rainforest beside creeks. There is an extensive track system — and a good place to begin exploring is West Head.

▲ Eastward lies **Barrenjoey Head**, its slopes covered in woodland and heath, and topped by a lighthouse. Barrenjoey is almost completely surrounded by ocean, except where joined to the mainland by the narrow sandy isthmus of Palm Beach.

West Head

West Head is at the northern end of the broad sandstone peninsula separating Pittwater on the east from Cowan Water on the west. Entry to this part of the Park is from Mona Vale Road at Terrey Hills. West Head is about 15 km through the Park.

There is a regular ferry service from Palm Beach to beaches on the eastern shore of West Head. A walking track leads from Mackeral Beach, via Resolute Beach to West Head.

▲ From West Head are some of the most spectacular views in Sydney — to the north across **Broken Bay** towards Gosford and the beaches, settlements and national parks of the Central Coast. At the centre, the 'crouching' shape of **Lion Island**, a Nature Reserve. General access to Lion Island is restricted so that wildlife — including Fairy Penguins — have a chance to live in peace.

▲ Framing these views are twisted branches of the Smooth-barked Apple, ***Angophora costata***, also known as Sydney Red Gum. This huge tree is kept in place by massive root growth amongst the rocks. On the right is a young Port Jackson Pine ***Callitris rhomboidea***.

Magnificent specimens of *Angophora costata* may be seen on the headlands around Broken Bay and the lower Hawkesbury River. *Callitris rhomboidea* is a native conifer in the cypress family, Cupressaceae. It grows on the hills around Broken Bay.

▲ At **Red Hands Cave** you will see hand stencils in red ochre, paint made from crushed iron-rich stone.

▲ The bay to the south of West Head Lookout is **Pittwater**, dotted with moored sailing boats; Sydney's suburbs reach to its southern and eastern shores. Eucalypt open-forest covers the sandstone hillslopes of the national park, while there are interesting occurrences of Spotted Gums, *Corymbia maculata*, on more shaley lower slopes of the opposite side of Pittwater in the distance.

▲ West Head has many reminders of the Aboriginal people who lived here prior to 1788. From near the lookout you can walk through the woodland to Red Hands Cave. Further back along West head Road, tracks lead to rock platforms with Aboriginal carvings.

▲ From the heights of West Head you can also walk down through sandstone hillside open-forest to Resolute Beach to cool off in the sea. Chains of Neptune's Necklace seaweed, the brown alga ***Hormosira banksii*** grow on the rocks and may be seen at low tide.

Along the West Head road

The West Head road passes for about 15 km along the sandstone ridgetops through extensive areas of **woodland** and **heath**, with occasional distant water glimpses beyond the white trunks of the Scribbly Gums. Marked walking tracks leading away from the main road enable you to enjoy this wonderful bushland and its coastal setting.

◄ The Spider Flowers — grey-flowered ***Grevillea buxifolia***, and red-flowered ***Grevillea speciosa*** — and the bright pink ***Boronia ledifolia***, are part of the brilliant display of spring colour.

The Basin Track

▲ The West Head road leads through the sandstone heath and woodland with white-trunked Scribbly Gums ***Eucalyptus haemastoma***. There is always something flowering here, but August–September is the time to see the greatest variety.

Typical of the West Head tracks is **The Basin Track** which begins about 4 km from West Head. Plants here include *Acacia ulicifolia, Bossiaea scolopendria, Zieria laevigata* and *Phebalium squamulosum*.

▲ About 500 metres or so along the track is **The Basin Aboriginal Engravings Site.** Here signs explain a number of different figures carved into flat sandstone rock surfaces. From this site the Aboriginal artists had wonderful 360° views. Please respect these carvings and do not walk on or touch them.

▲ Depressions in the rock between groups of carvings provide habitat for heath plants able to survive in very shallow soil. These form miniature 'bonsai gardens'. The mosses carpeting the shallow soil of these rock platform 'gardens' dry out intermittently, but become bright green again after rain.

The best times to visit these rock platforms are early morning and late afternoon, when low-angled sunlight highlights the lines of the Aboriginal carvings.

The track leads on about 2 km to The Basin, a large pool formed behind a sandy spit on the Pittwater shore. There are picnic and camping facilities here, and a regular ferry service to Palm Beach via Currawong and Great Mackeral Beaches.

◀ On the sheltered south-facing slope is open-forest, with a thicker understorey and different trees compared with the surrounding sandstone. Trees include *Livistona australis* Cabbage Palm.

Volcanic intrusion

Just before the West Head road ends at West Head, it crosses a **dyke of volcanic rock**. This is noticeable where the previously level road dips downwards, then up again. The volcanic rock is softer and more weathered than the surrounding sandstone, causing the dip. It also weathers to clay-rich soil that is more fertile than, and has different drainage characteristics from, the sandstone soil. As a result the vegetation is different.

▲ On the north-facing slope of this volcanic rock outcrop are Forest Oaks *Allocasuarina torulosa*, small trees with soft-ended needle-like 'leaves' that are really modified stems called 'cladodes'; these make a gentle sighing sound when the breeze catches them. They fall and carpet the ground — so you will not see so many shrubs here. Though they resemble pines in appearance, species of *Allocasuarina* are flowering plants in the Casuarinaceae family and not related to conifers.

▲ Prominent in the understorey are Burrawangs *Macrozamia communis*, plants that look palm-like, but are actually cycads — related more to conifers than palms; palms are flowering plants related to grasses and sedges. Burrawangs have separate male and female plants. Their large red seeds borne in cones are poisonous, and Aboriginal people treated them for several days before eating them.

◀ You may also find small *Synoum glandulosum* trees here, in the same family, Meliaceae, as Red and White Cedars. Their pinnate leaves have prominent glands along the mid-vein. The fragrant white autumn flowers develop into 3-lobed fruits containing shiny bronze seeds with bright orange 'arils' — food bodies that attract birds and animals that help disperse the seeds.

A peaceful, sheltered bay

Bobbin Head

Bobbin Head on Cowan Creek in the south-west corner of the Park has the main **Visitor and Information Centre**, with a Wildlife Shop selling maps and information about the park as well as more general nature guidebooks and publications.

Bobbin Head is accessible by car from North Turramurra along the Bobbin Head Road, or via the Ku-ring-gai Chase Road from Mt Colah. Public transport access to Bobbin Head is by train to Turramurra station then bus to the Park. There is also access to the western side of Ku-ring-gai Chase from railway stations at Cowan, Berowra and Mt Kuring-gai. Walking tracks lead down to bays along Cowan Creek and along the foreshores. A round trip between Berowra and Mt Kuring-gai is worthwhile. From Mt Kuring-gai station, Bobbin Head is a 5 km walk via Appletree Bay.

▲ At Bobbin Head the shores of Cowan Creek have been cleared to form picnic and parking areas. Here some of the local wildlife may come to watch you eat your picnic lunch. White Sulphur-crested Cockatoos and grey Wood Ducks are native species that have benefited from the changed conditions associated with urbanisation. Upstream a bridge leads to a boardwalk through the mangroves. Here you can see the trees and wildlife of this muddy habitat at close hand.

▲ Mangroves grow in mud and silt deposited in the tidal zone at the interface of fresh and salt water. The two Sydney species of mangrove are here — ***Avicennia marina***, the Grey Mangrove, has greyish-green, opposite-paired leaves and grows in saltier conditions, while ***Aegiceras corniculatum***, the River Mangrove, has brighter yellowish-green alternate leaves and grows on the landward side of the mangrove belt where water is fresher.

◀ Between the mangroves and the rocky shoreline is land flooded less often by the tides. Here grow small ***Casuarina glauca*** Swamp Oak trees, with rough bark often decorated with lichens. Beneath the Swamp Oaks is a ground cover of sedges that includes *Isolepis nodosa,* which carries its round seed heads attached to the sides of thin sharp stems.

▲ At the end of the boardwalk, a 15-minute track leads up past the sandstone outcrops through Sydney sandstone open-forest with trees of ***Eucalyptus piperita***, the Sydney Peppermint. The leaves have a strong 'eucalyptus' smell when crushed.

◄ *Petrophile pulchella* in the Proteaceae family grows well in shallow soil on rock ledges.

◄ The track leads past the 'grass skirts' of many ***Xanthorrhoea arborea*** Grass Trees to end in a narrow 'gallery' of rainforest confined to creek banks.

▲ A lookout at **Kalkari Visitor Centre** gives a good view over bush-clad ridgetops and rugged hillsides characteristic of the Hawkesbury Sandstone terrain.

▲ You may also notice Aboriginal **grinding grooves** on some of the rock shelves along here. These were used in the grinding process to sharpen tools such as axe heads and spear points, and are always close to water.

Kalkari provides displays and information about Ku-ring-gai Chase and its wildlife. Here there is a level nature trail with signs identifying bush plants and highlighting their interesting characteristics. Kalkari includes an interesting pond and is fenced, as it is also inhabited by kangaroos and emus. It is about 3 km westward along the Ku-ring-gai Chase Road, and is also accessible by walking tracks from Bobbin Head.

Bushland of Western Sydney is different!

Western Sydney's bushland is different from that of other parts of Sydney. It lacks the familiar colourful shrubs of the sandy soils — such as waratahs and boronias — and lacks the dominant topographic features of the sandstone landscapes. In their place is the gently undulating Cumberland plain with its Wianamatta Shale and alluvial soils, and a much lower rainfall. The understorey is predominantly grassy and herbaceous, and many of the plants are small and often only noticed when in flower. It is worth seeing however, and there are many ecological insights to be gained from a landscape that has affinities with the drier inland of New South Wales.

Plant communities include Cumberland Plain Woodland, Sydney Coastal River-flat Forest, Castlereagh Woodlands as well as small patches of Dry Rainforest. Most of these are listed as Endangered Ecological Communities under the *NSW Threatened Species Conservation Act.*

Scheyville National Park

Western Sydney has a number of relatively new National Parks and Nature Reserves.

North-east of Windsor is Scheyville National Park. Access is by car although you can take a train as far as Windsor. This is a new conservation area and includes land that was formerly rural. Woodland here has survived in areas that were set aside for military purposes and has a history of relatively low stock grazing.

Barking up the wrong tree!

The bark of eucalypt trees provides important clues to their identification. The common trees in western Sydney are readily distinguished by their bark.

◄ Forest Red Gum ***Eucalyptus tereticornis***, has grey to white smooth-barked trunk, and its thin bark peels off in flakes. Smooth-barked eucalypts are known as 'gums'. Other western Sydney gums are the Spotted Gum *Corymbia maculata*, found occasionally in Cumberland Plain woodland, *Eucalyptus deanei* and *Eucalyptus amplifolia* of the River-flat Forests, and *Eucalyptus sclerophylla* and *Eucalyptus parramattensis* of the Castlereagh Woodlands.

◄ Grey Box ***Eucalyptus moluccana*** has smooth-barked upswept branches but a fine fibrous rough bark on its trunk. This bark is known as 'box' bark. Other western Sydney boxes are Blue Box *Eucalyptus baueriana* and the rare Coast Grey Box *Eucalyptus bosistoana*, both associated with River-flat Forests.

► You can see Cumberland Plain Woodland and Castlereagh Woodlands in **Scheyville National Park**. Trees are being allowed to recolonise formerly cleared farmland areas but it will take many years for trees to reach a substantial size. Understorey plants such as Kangaroo Grass *Themeda australis* are also able to spread in the absence of grazing.

◀ Ironbarks stand out because of their thickly-furrowed black trunks. Narrow-leaved Ironbark, ***Eucalyptus crebra*** is common in Cumberland Plain Woodland.

◀ Stringybarks and mahoganies have persistent fibrous stringy bark that comes off in strands. Thin-leaved Stringybark ***Eucalyptus eugenioides*** is western Sydney's most common stringybark.

◀ ***Acacia pubescens*** was grown as a desirable garden plant in England as early as 1790, yet is now vulnerable in its natural western Sydney habitat.

◀ Alluvial creeklines are important habitats in western Sydney as most watercourses in rural areas have been cleared, channelled, or are weed infested. This creekline in Scheyville National Park is in relatively good condition.

▲ Scheyville National Park includes **Longneck Lagoon,** a substantial wetland with extensive open water, seen here during an extremely dry period in 1998. Periodic natural drying out stimulates germination of wetland plants.

Cattai National Park

For an interesting comparison with Longneck Lagoon visit Reedy Swamp in Cattai National Park (formerly Mitchell Park) which has a large area of periodically wet ground rather than open water. As a result there is much more herbaceous growth particularly of *Persicaria* species. There is also some interesting River-flat Forest with *Eucalyptus tereticornis*, as well as rainforest.

Windsor Downs Nature Reserve

Windsor Downs Nature Reserve at Bligh Park south of Windsor includes Cumberland Plain Woodland on shale soil influenced by Tertiary Alluvium with an intermixing of species from these two soil types.

◀ ***Grevillea juniperina*** is a rare spiky shrub that occurs naturally only in this part of western Sydney. Its red flowers hang from the ends of drooping branches.

Ironbarks and Scribbly Gums

Castlereagh Nature Reserve

The **Castlereagh Nature Reserve** at Llandilo, between Penrith and Windsor, is one of the best places to see Castlereagh Woodlands. Access is by car from the Northern Road where there is a small parking area. From here tracks lead across the Reserve. Slight changes in slope and differences in soils mean that you will see different groups of species as you walk through different parts of the bushland. The plants vary, depending on whether gravels, clays or sand are outcropping at the surface.

▲ Ironbarks, often blackened by fire, are the most distinctive trees, and occur where soils are clayey and gravelly. ***Eucalyptus fibrosa*** Broad-leaved Ironbark (right), is the most common species. Mugga Ironbark, ***Eucalyptus sideroxylon***, left, also occurs here. A pink-flowered form is popular for garden and street plantings.

In September the yellow flowers of pea family shrubs, often referred to broadly as 'bacon and eggs' because of purple or brown markings, are abundant in the understorey beneath the ironbarks.

▲ Here is the trailing broad-leaved ***Podolobium scandens***. Other yellow pea-flowered shrubs here are *Dillwynia tenuifolia, Pultenaea parviflora, Daviesia ulicifolia, Daviesia acicularis* and *Pultenaea villosa.*

◄ A rare local endemic sheoak is ***Allocasuarina glareicola*** which is restricted to a few small populations in Castlereagh Nature Reserve. The female flowers are shown here.

◄ Local populations of the small cycad ***Macrozamia spiralis*** may be seen. The poisonous seeds were eaten by Aboriginal people after they had treated them to make them safe to eat.

◀ Yellow highlights also come from *Acacia* flowers, in August–September the brilliant yellow of ***Acacia elongata*** and in May–June the paler cream of *Acacia falcata*.

◀ Castlereagh Swamp Woodland with *Eucalyptus parramattensis* and paperbarks *Melaleuca decora* occurs in poorly-drained sites. In September you may see mauve flowers of donkey orchids ***Diuris punctata*** amongst the sedges *Cyathochaeta diandra* and *Ptilothrix deusta*, and sundews, species of *Drosera*.

▲ Where the soil is sandier, trees are smaller than on the clayey soils. 'Graffiti' made by moth larvae can be seen on the smooth white and grey trunks of the Hard-leaved Scribbly Gum ***Eucalyptus sclerophylla. Banksia spinulosa*** flowers (foreground, right) provide food for birds and small mammal pollinators in autumn.

For a surprise!

While in the area visit **Agnes Banks Nature Reserve.** Access is from Rickards Road, Agnes Banks, south of Richmond. People are always surprised to see the banksias — *Banksia aemula, Banksia serrata, Banksia oblongifolia* and *Banksia spinulosa* — and other coastal species that grow on the small remnant of white leached sand. A series of sand dunes, swales and swamps once supported a fascinating variety of local vegetation types — most have been destroyed by sand extraction but what remains still provides a most 'un-Western Sydney-like' landscape.

▲ ***Philotheca salsolifolia*** in the Rutaceae is a frequent shrub on the sand here.

Wildflowers in the south-west

If you're interested in the plants of western Sydney or indeed Australian plants in general head down to **Mount Annan Botanic Garden** 57 km south-west of Sydney, just off the F5 Freeway on Tourist Drive 18 between Campbelltown and Camden. Buses run between Campbelltown and Camden. Mount Annan Botanic Garden, on former farmland, has been developed by the Royal Botanic Gardens Sydney to feature Australian plants. The Garden, in a 400 ha setting, was opened in 1988. A sort of 'botanical equivalent' of the Western Plains Zoo!

Feature gardens include the Terrace Garden, presenting a wide range of Australian plants in their family groupings, from ferns, cycads and conifers, through to the more advanced flowering plants, and showing the great variety of plant forms and lifestyles that have evolved.

Roads and paths lead to other displays of major Australian plant groups, such as the Bottlebrush Garden, Wattle Garden and Banksia Garden, and the Eucalypt Arboretum and Fig Arboretum. There is a Sundial of Human Involvement and a children's playground. And the picnic facilities are very good.

▲ **The Garden Shop & Visitor Centre** near the entrance has a range of botanical books and information. Around the Visitor Centre is a garden displaying some of the colourful native species of western Sydney, and rare plants from other parts.

From here you may walk to areas where remnants of the original bushland of this part of Sydney survived in the former farmland. They are now being regenerated as significant features of the Botanic Garden, and include the Woodland Conservation Area and Mount Annan summit.

▲ The **Woodland Conservation Area,** a remnant of the Cumberland Plain Woodland that once covered most of western Sydney, is a ten minute stroll from the Visitor Centre. Here you can walk through the woodland along a mown track, and see close at hand its characteristic trees. The Grey Box ***Eucalyptus moluccana*** (foreground, centre), is the most frequent tree in the Woodland Conservation Area. It has fibrous rough bark on the trunk and smooth upswept branches. Also present is Forest Red Gum ***Eucalyptus tereticornis*** (right), and here and there Narrow-Leaved Ironbarks *Eucalyptus crebra* stand out because of their thickly-furrowed black trunks. ***Bursaria spinosa***, the most abundant shrub, provides a display of white fragrant flowers in summer and early autumn.

Mount Annan Botanic Garden was grazing land before becoming a botanic garden.
In the Woodland Conservation Area native understorey plants are now regrowing.
Marked plots in the woodland are part of a monitoring project by scientists of the Royal Botanic Gardens.

◄ *Themeda australis* Kangaroo Grass, is the most common of the native grasses in the woodland. Its decorative seed heads in metre-high tussocks give the woodland a distinctive red-brown sheen during the warmer months.

◄ The worldwide bluebell family Campanulaceae is represented in woodland at Mount Annan Botanic Garden by ***Wahlenbergia communis*** and several other species.

▲ In spring the woodland ground cover is brought to life by the flowers of daisies, orchids, lilies and other small herbs.

▲ From the **summit of Mt Annan,** a pleasant walk south from the Visitor Centre, there are spectacular views over the Botanic Garden and surrounding countryside. On a clear day you can see Sydney's CBD, with its distinctive Centrepoint Tower.

◄ The woodland harbours a dense patch of the daisy ***Rhodanthe anthemoides***, now rare in western Sydney.

◄ Also present in the woodland is the endangered ***Pimelea spicata***, a small trailing herb, inconspicuous until it flowers. It belongs in the Thymelaeaceae family.

◄ Clumps of Yellow Burr-daisy ***Calotis lappulacea***, a local native colonising species, fringe the Mt Annan summit track. A remnant of Dry Rainforest is being regenerated on the summit and you can see where thickets of the woody weed African Olive *Olea europaea* subspecies *africana* have been cleared. Originally introduced as a hedge plant in the nineteenth century, it has naturalised on steeper slopes in the area. Its seeds are dispersed by birds and you will often see thickets of young plants coming up under remnant eucalypts.

Introducing the Blue Mountains

The Blue Mountains to the west of Sydney is a massive sandstone plateau rising gradually to a height of just over 1000 m. Some of its highest peaks are the basalt caps of Mt Tomah and Mt Wilson. The upper Blue Mountains between Wentworth Falls and Mt Victoria, between 900 and 1100 m, has the highest rainfall and receives light winter snowfalls. The sandstone plateau is dissected by entrenched rivers which ultimately all flow to the Hawkesbury River and have cut deep gorges through the sandstone. The Coxs River flows from Lithgow to join the Nepean River south of Penrith, and has been dammed at Warragamba just above this junction to form Lake Burragorang, Sydney's main water supply. The Grose River flows eastward from Blackheath to the Nepean near Richmond while the Wollangambe flows from the north-west to join the Colo and then the Hawkesbury.

Two roads cross the Blue Mountains and lead to the west. These follow the main ridges between the river catchments. The Great Western Highway crosses south of the Grose River, via Penrith, Glenbrook, Katoomba, Blackheath, Mt Victoria and Lithgow. Bells Line of Road crosses north of the Grose, via Richmond, Mt Tomah, Bell and Lithgow. A connecting road between Bell and Mt Victoria allows an easy round trip.

Blue Mountains National Park covers nearly 250 000 ha and its northern section, the Wollangambe Wilderness, joins the larger Wollemi National Park to make up one of the largest wilderness areas in New South Wales. Here you can explore and experience Sydney bushland in a setting quite remote from suburbs and development, at time scales ranging from a fifteen minute walk to a hike of several days. However, the Park is not without human impact — weeds can be found in some places, particularly around the edges, and some former camping areas and popular tourist sites have been loved to their detriment. Nevertheless, the rich diversity of plant and animal life in largely natural condition has led to the area's nomination for World Heritage status. Much of this natural diversity can be enjoyed on relatively short walks.

The lower Blue Mountains

Glenbrook just west of Penrith is a good place to begin exploring the Lower Blue Mountains. The Blue Mountains National Park **Visitor Information Centre** is a 2 km drive from the Great Western Highway or a 1 km walk from Glenbrook railway station. At the Visitor Centre, signs explain local geology, flora and fauna and indicate walking tracks to a number of places including Glenbrook Gorge. Most tracks traverse sandstone, the predominant rock, and pass through its characteristic vegetation. From here there is also car access to picnic sites such as The Ironbarks — ironbark forest on a remnant shale cap — and Euroka — a volcanic area that has been mostly cleared although there are still some remaining Blue Gums. Other areas that are also accessible are Murphys Glen and Tobys Glen. These give views of the tall forest growth possible on richer volcanic soil; *Eucalyptus deanei* Deane's Gum trees at Tobys Glen are among the tallest trees in the park.

Nepean River via Glenbrook Gorge

The spectacular gorge-like valley of Glenbrook Creek is easily accessible from the **Visitor Information Centre.**

The walking track begins in sandstone ridgetop woodland, quickly reaches a steep sheltered sandstone slope with open-forest, and then descends to bouldery Glenbrook Creek with its sandstone riparian scrub. From here you can look up to woodland and open-forest on exposed sandstone hillslopes.

▲ In September, in the open woodland understorey near the Glenbrook Visitor Information Centre, the bright yellow and brown pea flowers of ***Bossiaea obcordata*** may be prolific.

▲ The track to Glenbrook Creek winds across the sandstone ridge under trees of Narrow-leaved Apple *Angophora bakeri*, their fibrous-barked trunks contrasting with the tessellated rough bark of Red Bloodwood *Corymbia gummifera*. Amongst the shrubs is the Large Wedge Pea ***Gompholobium grandiflorum***.

◄ As you approach the edge of the gorge, the smooth, pink to grey, dimpled trunks of ***Angophora costata*** trees — the Sydney Red Gum or Smooth-barked Apple — become more noticeable, and long spiky leaves of Grass Trees ***Xanthorrhoea arborea*** stand out beside the track.

▲ The track now descends the steep and rugged but sheltered hillside. Here woodland changes to open-forest, with characteristic Sydney Peppermint ***Eucalyptus piperita*** trees. There are also some Grey Gum *Eucalyptus punctata* and Turpentine trees *Syncarpia glomulifera*, indicating moister more fertile conditions than on the ridgetop.

► Small trees, shrubs and ferns grow well in the shelter near the base of the slope. ***Leionema dentatum*** (formerly *Phebalium dentatum*), is a soft-leaved shrub in the Rutaceae or citrus family. Its cream flowers are out in spring.

The full walk to the Nepean River is a fairly strenuous halfday, with much rock-hopping and creek crossing. However, it can be cut short at any time after reaching the creek and yet still allow you to see the sandstone flora from ridgetop woodland to riparian scrub.

Rock-hopping along Glenbrook Creek

Glenbrook Creek occupies a spectacular gorge-like valley, between steep hillsides with trees among sculpted rocky outcrops, and the creek flowing across and occasionally under large slabs of sandstone. Amongst the rocky boulders, in and along the creek, is riparian scrub with small trees and large shrubs — Water Gum *Tristaniopsis laurina*, and Teatree *Leptospermum polygalifolium* in the Myrtaceae family; and River Lomatia *Lomatia myricoides* and Scrub Beefwood *Stenocarpus salignus* in the Proteaceae family– all leaning downstream, bent over by periodic floods.

▲ Unusual features of the creekside riparian scrub here are several uncommon species including the rare shrub ***Grevillea sericea* subspecies *riparia*** which is a restricted to rocky banks of Blue Mountains rivers such as the Colo and Grose, and here. Its beautiful pink flowers appear in September.

▼ White-flowered fragrant ***Philotheca myoporoides*** (previously *Eriostemon myoporoides*) has long been a popular horticultural species, and you are likely to find it in many gardens. It's a thrill to find it growing beside the creek here in its natural habitat.

The creek's undisturbed banks are a welcome change from sandstone creek banks nearer Sydney that are made weedy by closer suburban development.

▲ Shrubs of ***Daviesia corymbosa*** grow occasionally along the rocky creek bank, with clusters of bright yellow and brown flowers in spring.

▲ In backwater pools along Glenbrook Creek trailing leaves of aquatic plants such as ***Triglochin*** Water Ribbons can be seen in the more slowly moving water. Soft fronds of ***Calochlaena dubia***, sometimes called False Bracken, droop over the water from these more sheltered banks.

▲ From the creek banks look up to spectacular rock faces on the opposite exposed hillside, glinting orange in sunlight. Twisted branches are likely to belong to Yellow Bloodwoods ***Corymbia eximia*** clinging among the rocks — their large cream flowers burst forth in October. Bluish leaves belong to ***Eucalyptus agglomerata***, the Blue-leaved Stringybark.

When rock-hopping along the creek around big sandstone boulders and across small sandy beaches you need to take care to avoid slipping on rocks. Passengers on trains that pass by regularly halfway up the steep slope above the creek have an easier but more distant view.

At the junction with the Nepean River is a massive rockpile, said to be partly construction material from the Glenbrook Railway tunnel. River Oaks *Casuarina cunninghamiana* colonising here are the signature of the Nepean River banks.

A cool climate mountain garden set in a sandstone wilderness

Mount Tomah Botanic Garden in the Blue Mountains is the cool climate garden of the Royal Botanic Gardens Sydney. It is on Bells Line of Road, 12 km west of Bilpin and 105 km west of Sydney. There is a **Visitor Centre** with a bookshop and restaurant.

▲ The garden area has been developed to feature plantings of southern hemisphere and native high altitude plants and also includes conifer and rhododendron collections, a Gondwana Walk, a Rock Garden, a Formal Garden, Rainforest Walk and viewing platform, Plant Explorers Walk and spectacular spring and autumn displays.

▲ At 1000 m altitude, **Mount Tomah Botanic Garden** is perched on one of the Blue Mountains' basalt caps, just north of the gorge of the Grose River and surrounded by rugged sandstone country. The fertile volcanic soils of Mt Tomah contrast with the infertile sandy soils of the sandstone landscape, and were cleared for farming many years ago.

◀ Magnificent trees of Brown Barrel ***Eucalyptus fastigata,*** up to 40 metres high, once soared above the treefern understorey of tall eucalypt forest on the basalt soils. Most of the forest was cleared to make way for the first farms.

The Mount Tomah Botanic Garden opened to the public in 1987. It was a former cut-flower farm donated to the people of New South Wales by Effie and Alfred Brunet.

▲ Visitors to Mount Tomah Botanic Garden can picnic beneath some remaining Brown Barrel trees, though the understorey is now more likely to be garden beds of rhododendrons.

▲ The **Gondwana Walk** passes through warm temperate rainforest on a sheltered south-facing slope, with treeferns, Coachwoods *Ceratopetalum apetalum* and Sassafras trees *Doryphora sassafras*.

Take a walk on the wild side away from the tourist crowd

Mount Tomah Botanic Garden also includes an area of sandstone plateau vegetation as well as rainforest patches growing in the sheltered valleys facing east and south.

There are expansive views across the surrounding sandstone landscape from the Garden. **Ecotours** into the sandstone landscape of Mount Tomah Botanic Garden can be arranged for those who want to see a variety of plants and habitats.

◄ *Alania enlicheri* is a lily-like plant that has a very specialised sandstone habitat. It grows only on moist rock faces or in cliff-line shrub communities. It is very restricted and occurs naturally only in the Blue Mountains and Hornsby Plateau areas. This is the only species in the genus *Alania*. It belongs to the Anthericaceae family, formerly part of the lily family.

Low **heath** on level ridgetops of the sandstone landscape contrasts with forest cover on distant rounded basalt caps.

◄ This Triggerplant ***Stylidium productum*** is common in the sandstone woodland. Gentle poking at it with a fine grass stem will trigger the flower to respond, then it will gradually reset.

▲ The bright yellow flowers of the Sunshine Wattle ***Acacia terminalis***, have attracted the attention of this group of people enjoying the bush on an Ecotour. 'Oh lovely is the Wattle, the emblem of our land. You can stick it in a bottle, or hold it in your hand.'

The upper Blue Mountains experience

No visit to the Blue Mountains is complete without seeing the misty heights of the upper Mountains. Here the main sandstone plateau is between 900 and 1100 m, has a rainfall of 1400 mm per year and occasionally receives light snowfalls.

There is easy access to bushland, walks and viewing points from the upper Blue Mountains towns of Wentworth Falls, Leura, Katoomba, Medlow Bath, Blackheath and Mt Victoria. Access from railway stations is either an easy walk or a short bus ride.

The upper Blue Mountains also has a wide range of accommodation and eating establishments for *après le bushwalk*.

▲ The iconistic profile of the **Three Sisters** is perhaps almost as well known as that of the Opera House. Their home near Echo Point at Katoomba is well marked on the tourist trail.

▲ From Echo Point a short well-made path leads to the Three Sisters. This takes you along the ridge with extensive views through heath and woodland. Tall, smooth-barked trees of Blue Mountain Ash ***Eucalyptus oreades*** grow on moist sheltered places. This species is one of the few eucalypts that is killed by fire. It regrows quickly from seed but you will only see big trees in sites that have not been burnt for a long time.

◀ At the Three Sisters however the path becomes a staircase. There are 1000 steps down to where rainforest, with trees of Coachwood *Ceratopetalum apetalum* and Sassafras *Doryphora sassafras*, shrubs of *Callicoma serratifolia* and *Tristaniopsis collina* and treeferns, thrives on the moist sheltered slopes below the waterfalls.

Federal Pass takes you around the base of the cliffs to the old coal mines, and further on to the Landslide and Ruined Castle. The Scenic Railway, once used by coal miners, provides an easy way up to the top of the ridge, though there are also steps for the hardy.

▲ **Rainforest** is restricted to sheltered sites with plenty of moisture. The bright green of the rainforest canopy contrasts with the yellow-green of the eucalypt forest and woodland here just below the Scenic Railway terminus.

From Echo Point there are also paths along the top of the cliffs (for example, the Prince Henry Clifftop Walk) leading to Katoomba Cascades and Katoomba Falls, the Scenic Skyway and Scenic Railway. The paths provide an easier descent into the valley, or in the other direction, to Leura Cascades.

◄ Sunny spots alternate with cool shady glens along many tracks. One of the pleasures of the upper Mountains walks are the frequent meetings with falling water and cascading streams. Unfortunately urban development in its catchment has reduced the water quality and allowed weeds to invade the bush.

◄ Not all of the bushland tracks are dogged with steps. Provision is now made for some wheelchair access tracks such as here near Katoomba Cascades.

There are many other bushland places worth visiting in the upper Blue Mountains. At **Blackheath** there is a National Parks and Wildlife Service **Visitor Information Centre** near Govetts Leap. There is a short education walk here. From Govetts Leap Lookout there are splendid views over the gorge of the upper Grose River to Mount Banks, a basalt capped mountain on the northern side of the Grose. From here tracks lead along the clifftops through heath, sedgeland and woodland or down into the moist forest of the valley. An interesting shorter walk leads from Evans Lookout through the Grand Canyon, a deep sandstone rainforested ravine on Greaves Creek.

To find out more

We trust you have enjoyed your journey through some of Sydney's bushland with us. If you have the opportunity, we know you will also enjoy discovering more for yourself. There are many things we have not been able to cover in this book — for example the coastal scenery and rainforest of the Illawarra, the heath, woodlands and beaches of the Central Coast, and the moist forests of the Southern Highlands and Kangaroo Valley.

And remember, we still know very little about the complex biology and ecology of most of our plants. Even new species are still being found — from large trees like the Wollemi Pine discovered in 1994, to microscopic fungi growing with tree roots. By observing and asking questions you may have the chance to make an exciting new scientific discovery yourself!

Here are some ways to help you start finding out more — not a comprehensive list, but each will lead you to further sources of information.

Landscape and vegetation of Sydney's bushland

Taken for Granted: the bushland of Sydney and its suburbs by Doug Benson & Jocelyn Howell (1995) Kangaroo Press, Kenthurst.

Mountain Devil to Mangrove: A guide to natural vegetation in the Hawkesbury-Nepean catchment by Doug Benson, Jocelyn Howell, & Lyn McDougall (1996) Royal Botanic Gardens, Sydney.

Missing Jigsaw Pieces: The Bushplants of the Cooks River Valley by Doug Benson, Danie Ondinea & Virginia Bear (1999) Royal Botanic Gardens, Sydney.

Rare Bushland Plants of Western Sydney by Teresa James, Lyn McDougall, & Doug Benson (1999) Revised Edition, Royal Botanic Gardens, Sydney.

Cunninghamia: a journal of plant ecology for eastern Australia includes papers dealing with all aspects of plant ecology in particular vegetation mapping and plant community dynamics and management. Published twice a year by the Royal Botanic Gardens, Sydney. Available on subscription or as individual issues available separately from the Royal Botanic Gardens Shops, phone (02) 9231 8125.

Vegetation maps available from Royal Botanic Gardens Shops:

Natural vegetation of the Sydney area by Doug Benson & Jocelyn Howell (1994) 1:100 000 vegetation map covering eastern Sydney from Broken Bay to Botany Bay, together with 1:40 000 map of Vegetation of Ku-ring-gai Chase National Park and Muogamarra Nature Reserve. Originally published in *Cunninghamia* Volume 3 Number 4.

Natural vegetation of the Penrith area by Doug Benson (1992) 1:100 000 vegetation map covering western Sydney and the lower Blue Mountains.

Natural vegetation of the Katoomba 1:100 000 map sheet by David Keith and Doug Benson (1988). Available in *Cunninghamia* Volume 2 Number 1.

Plant identification

Field Guide to the orchids of New South Wales and Victoria by Anthony Bishop (1996) UNSW Press, Sydney.

Field Guide to Eucalypts Volume 1: South-eastern Australia by M.I.H. Brooker & D.A.. Kleinig (2nd edition 1999) Bloomings Books, Hawthorn, Victoria.

Common weeds of Sydney bushland by Robin Buchanan (1981) Inkata Press, Melbourne.

Beach plants of south eastern Australia by Roger Carolin and Peter Clarke (1991) Sainty & Associates, Potts Point.

Flora of the Sydney region by R.C. Carolin & M.D. Tindale (4th edition 1993) Reed, Chatswood.

Sydney flora: A beginner's guide to native plants by Tony Edmonds & Joan Webb (2nd edition 1998) Surrey Beatty & Sons Pty Ltd, Chipping Norton.

Native Plants of the Sydney District by Alan Fairley & Philip Moore (1989) Kangaroo Press & The Society for Growing Australian Plants, Sydney.

Flora of New South Wales Volumes 1–4 edited by Gwen Harden (1990–93) plus Supplement to Volume 1 (2000) UNSW Press, Sydney.

Proteaceae of New South Wales edited by G.J. Harden, D.W. Hardin & D.C. Godden (2000) UNSW Press, Sydney.

Carnivorous plants of Australia Volume 3 by Allen Lowrie (1998) University of Western Australia Press, Nedlands.

Field guide to the native plants of Sydney by Les Robinson (1991) Kangaroo Press, Kenthurst.

Waterplants of Australia by G.R. Sainty & S.W.L. Jacobs (3rd Edition 1994). CSIRO Australia Sainty & Associates, Darlinghurst.

Grasses of New South Wales by D.J.B. Wheeler, S.W.L. Jacobs & B.E. Norton (2nd edition 1990) University of New England Monographs 3.

Trees and shrubs in rainforests of New South Wales and southern Queensland by J.B. Williams G.J. Harden & W.J.F. McDonald (1984) Botany Dept., University of New England.

Useful Websites: Royal Botanic Gardens Sydney: www.rbgsyd.nsw.gov.au
Australian National Botanic Gardens: www.anbg.gov.au/anbg/
Association of Societies for Growing Australian Plants: farrer.riv.csu.edu.au/ASGAP/

Bushland ecology, plant physiology and prehistory

Plants in action: Adaptation in nature, performance in cultivation edited by Brian Atwell, Paul Kriedeman & Colin Turnbull (1999) Macmillan Education Australia Pty. Ltd., South Yarra. Comprehensive text on plant physiology relevant to Australian conditions.

Ecology of Sydney plant species by Doug Benson & Lyn McDougall (1993–1999) Series published as Parts in *Cunninghamia: a journal of plant ecology for eastern Australia.*

Ant-plant interactions in Australia edited by Ralf C. Buckley (1982) Geobotany 4, Dr W Junk Publishers, The Hague.

Bush regeneration: recovering Australian landscapes by Robin Buchanan (1989) TAFE, Sydney.

Bibliography of fire ecology in Australia by Malcolm Gill, Peter Moore & Warren Martin (4th edition 1994) plus Supplement (1996) NSW National Parks & Wildlife Service, Hurstville.

Riverside plants of the Hawkesbury-Nepean by Jocelyn Howell, Lyn McDougall & Doug Benson (1995) Royal Botanic Gardens, Sydney.

Bush foods of New South Wales — a botanic record and an Aboriginal oral history by Kathy Stewart & Bob Percival (1997) Royal Botanic Gardens, Sydney.

The greening of Gondwana: The 400 million year history of Australia's plants by Mary E. White (3rd edition 1998) Kangaroo Press/Simon & Schuster Australia, East Roseville.

After the greening: The browning of Australia by Mary E. White (1994) Kangaroo Press, Kenthurst.

Winning the war against weeds by Mark A. Wolff (1999) Kangaroo Press/Simon & Schuster Australia, East Roseville.

Out and about in Sydney's bush

Those keen to enjoy more bushwalks can choose from a selection of books or brochures giving details about walks around Sydney, available in the Royal Botanic Gardens Shops at Sydney, Mount Annan and Mount Tomah, and at Visitor and Information Centres of the NSW National Parks and Wildlife Service.

Pick up the most recent guide to national parks throughout the state at NSW National Parks & Wildlife Service Information Centres — e.g. in the Sydney CBD at 102 George Street, The Rocks, or at head office, Level 1, 43 Bridge Steet, Hurstville. The NPWS also provides information about rare plants and endangered ecologcal communities. Also in the Sydney CBD is The Sydney Harbour National Park Information Centre, at Cadmans Cottage, 110 George Street, The Rocks. You can phone the NPWS Information Line, 1300 36 1967, or visit their website at www.npws.nsw.gov.au.

The voluntary organisation National Parks Association of NSW conducts non-commercial guided bushwalks for members and has a range of information and bushwalking books at Level 9, 91 York Street, Sydney, phone 9299 0000, website www.speednet.com.au/~abarca/. The Nature Conservation Council of NSW is the non-profit umbrella organisation for over 100 environment groups and scientific societies, and has a range of information at Level 5, 362 Kent Street, Sydney, phone 9279 2466, and at www.nccnsw.org.au.

You can get access to information on bus, train and ferry routes and timetables from the CityRail, Sydney Ferries and Sydney Buses InfoLine phone 131 500 and from State Transit's website www.131500.com.au.

Now that you've been introduced to our Sydney bushland, it's over to you to explore further!

Index to names of plants (illustrated on pages in bold type) and topics